# VII International Conference for Young Scientists "High Technology: Research and Applications" (HTRA 2018)

IOP Conference Series: Materials Science and Engineering
Volume 510



Tomsk, Russia
26 – 30 November 2018

ISBN: 978-1-5108-8631-5
ISSN: 1757-8981

**Printed from e-media with permission by:**

Curran Associates, Inc.
57 Morehouse Lane
Red Hook, NY 12571

**Some format issues inherent in the e-media version may also appear in this print version.**



Printed by Curran Associates, Inc. (2019)

For permission requests, please contact the Institute of Physics at the address below.

Institute of Physics
Dirac House, Temple Back
Bristol BS1 6BE UK

Phone: 44 1 17 929 7481
Fax: 44 1 17 920 0979

techtracking@iop.org

**Additional copies of this publication are available from:**

Curran Associates, Inc.
57 Morehouse Lane
Red Hook, NY 12571 USA
Phone: 845-758-0400
Fax: 845-758-2633
Email: curran@proceedings.com
Web: www.proceedings.com

# Table of contents

## Volume 510

**The VII International Conference for Young Scientists "High Technology: Research and Applications"**

**26–30 November 2018, Tomsk, Russian Federation**

**Accepted papers received: 27 March 2019**
**Published online: 24 April 2019**

### Preface



### Papers

IOP Conf. Series: Materials Science and Engineering **510** (2019) 011001 doi:10.1088/1757-899X/510/1/011001

# Preface

The VII$^{th}$ International Conference for young scientists "High Technology: Research and Applications 2018" (HTRA-2018) was held at Tomsk Polytechnic University, Tomsk. Russia on November 26-30, 2018.

The goal of the conference was the development of cooperation between young researchers in the field of engineering sciences and high technology application. Conference focuses on modern technology and the advances in science and technology based on an interdisciplinary approach to problem solving. The famous scientists and the heads of leading scientific schools shared their experience in industrial implementation of their research results, as well as the creation of small innovative enterprises.

The mains topics of the conference were as follows:

- Chemical technology and biotechnology
- Rational use of natural and technogeneous mineral raw materials and water resources
- Photonic technologies and photo engineering
- Reliability of mechanical engineering and mechanic technologies

On behalf of Organizing Committee of HTRA-2018, we would like to thank the sponsors of the conference - RFBR, as well as all delegates, guests and keynote speakers for their contribution to the success of the conference.

*Organizing Committee Chairman*

*Dr. A.N. Yakovlev*

# Peer review statement

All papers published in this volume of *IOP Conference Series: Materials Science and Engineering* have been peer reviewed through processes administered by the proceedings Editors. Reviews were conducted by expert referees to the professional and scientific standards expected of a proceedings journal published by IOP Publishing.

IOP Conf. Series: Materials Science and Engineering **510** (2019) 012001 doi:10.1088/1757-899X/510/1/012001

# Experimental investigation of wear resistance of copper coated electrode-tool during electrical discharge machining

**T Ablyaz [1], K Muratov[1], S B Preetkanwal[2], S S Sarabjeet[2]**
[1] Perm National Research Polytechnic University, Perm-614000, Russian Federation
[2] Mechanical Engineering Department, Beant College of Engineering & Technology, Gurdaspur-143521, India

*E-mail: lowrider11-13-11@mail.ru

**Abstract.** In this study, electrode-tools with excellent electrical and thermal properties were manufactured for use in die-sinker Electrical Discharge Machining. The most common materials, nowadays, for the manufacture of electrode tools are copper and graphite. The traditional method of obtaining complex profile electrodes tools is multi-axis metal cutting blade processing. The low utilization rate of the tool material during the shaping of complex-profile electrodes significantly affects the cost of a suitable product. The paper proposed a method of fabricating electrodes obtained by casting methods with the subsequent application of electro-erosion resistant coatings. The study of the investigating the durability of the coated electrode during the processing of 34CrNiMo6 steel has been carried out. It has been established that to obtain a uniform copper coating with a thickness of 300 μm, it is necessary to use a current density in the course of galvanic deposition of 1.3 A / mm² at an exposure of 5 hours. An experimental study of the durability of the coating during the die-sinker Electrical Discharge Machining of 34CrNiMo6 steel showed that with an increase in the current intensity above 8A, the copper coating burns out. Based on the research, it was found that the processing of 34CrNiMo6 steel with electrode having copper coating on aluminum alloy AK12 as a base metal showcased superior processing capacity. The economic efficiency of using coated electrodes was enhanced by 35% as compared to the technology for obtaining a tool from a solid copper billet.

## 1. Introduction

The technology of die-sinker electrical discharge machining (EDM) allows manufacturing complex shape parts made of materials with enhanced physiomechanical properties [1-5]. The main limiting factor in the use of die-sinker EDM machines is the limited technological capabilities in manufacturing complex profile electrode tools (ET) [2]. A majority of the total time in process of manufacturing parts by the EDM method is consumed while fabrication of desired ETs. The EDM utilizes ETs made up of materials with high thermo-electric and erosion resistant properties. Amongst these materials, copper and graphite are the first choice for ET fabrication.

Electrodes tools (ET) are fabricated in various ways; the most common methods are casting, turning, pressing, and processing by the method of selective laser melting and sintering (SLM). In EDM two types of ETs are categorized: profiled and non-profiled. The Profile tools are applicable in the die-sinker EDM wherein, the shape of the finished part corresponds to the tool profile. Aluminum, copper, and bronze are the most common electrodes used therein. Non-profiled tools are employed in wire-cut EDM equipment. Brass or copper wire is the tool that performs cutting on the end surfaces.

In the traditional manufacturing of ET having complex profile, its shape is divided into basic areas, and each section is fabricated separately. The individual elements of the electrode are then coupled in one tool. At present, metal-cutting techniques for producing ET of a complex profile are widely used. However, the final complex profile electrodes, thus produced, lack the accuracy while used in machining. As the cost of manufacturing a complex profile electrode depends on the complexity of the tool profile, often the processing of promising parts such an electrode becomes economically unprofitable [2, 3].

One of the ways to reduce the cost of manufacturing ET is to obtain complex ET with erosion-resistant coatings (cortical electrodes) [5]. Both metallic and polymeric materials can be used as a base for coating. The metal or polymer can be used as base material for manufacturing the complex ET employing Casting techniques and rapid prototyping technologies.

A similar research work [6] has highlighted the effective use of complex aluminum ET with copper coating for EDM of alloyed steel (figure 1).

**Figure 1.** Aluminum ET with copper coating:
1-aluminum base; 2-copper coating

It was found that the amount of wear of the complex ET from aluminum with copper coated is comparable to the copper electrode. Comparison of the performance of copper and complex electrode-tools, showed a difference of not more than 4% percent.

However, in order to fully understand the operational characteristics of the coated electrodes, this work is aimed to study the durability of the coated ET while ED machining of 34CrNiMo6 steel.

## 2. Materials and Methods

34CrNiMo6 steel was chosen as workpiece for carrying out the EDM experiments and Electronica make Smart CNC EDM machine was utilized for this purpose incorporating dielectric oil as working fluid.

**Table 1.** Materials for the preparation of galvanic solution

| Materials | **Concentration (g/l)** |
|---|---|
| Sulphuric acid | 30 |
| Alcohol | 5 |
| Distilled water | 1 l |
| Copper sulfate | 200 |

This study was conducted on three cylindrical samples of aluminum AL1 (figure 2).

IOP Conf. Series: Materials Science and Engineering **510** (2019) 012001 doi:10.1088/1757-899X/510/1/012001

**sample 1** **sample 2** **sample 3**

**Figure 2.** Copper-coated EI samples

The exposure time of each sample was 1 hour (table 2).

**Table 2.** Time and current strength in the experiment

| Sample | Current, A | Holding time, h | Current density, A/mm$^2$ |
|---|---|---|---|
| 1 | 3 | 1 | 0.95 |
| 2 | 4 | 1 | 1.27 |
| 3 | 2 | 1 | 0.63 |

After each exposure, the thickness of the coating was measured using an Infinity microscope at 100x magnification. After coating the samples, experimental studies of the electrical discharge machining of a billet of steel 34CrNiMo6 were carried out to measure the amount of wear for the applied cortical coating. Processing modes are presented in table 3. Each experiment was repeated three times.

**Table 3.** Processing parameters on the copy-piercing equipment

| Mode | Current, A | Ton, MKS | Voltage, V |
|---|---|---|---|
| Max | 8 | 150 | 100 |
| Min | 2 | 150 | 100 |
| EDM time - 20 minutes | | | |

## 3. Results and discussion

The results of the thickness of the copper coating as per the parameters (table 2) of galvanic deposition are presented in table 4.

**Table 4.** Formation of the thickness of the coating samples after an hour of exposure

| Sample | Current, A | Coating thickness, micron |
|---|---|---|
| 1 | 3 | 76.98 |
| 2 | 4 | 112.12 |
| 3 | 2 | 42.23 |

The results of the study showed that aluminum samples 1 and 3 with an exposure of 1 hour (table 2) have a small coating thickness, which will certainly lead to its burning out upon contact with workpiece while EDM. Sample 2 of aluminum (figure 3) witnessed the highest coating thickness among all samples after copper plating and holding for 1 hour.

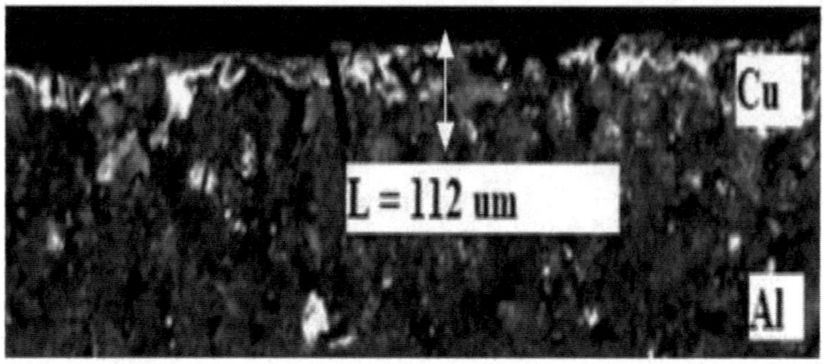


**Figure 3.** The thickness of the copper coating sample 2

After obtaining the suitable parameters for galvanic deposition by current values, the effect of exposure time on the formation of the coating thickness with a step of 1 hour was studied. It has been established that to obtain a uniform copper coating with a thickness of 300 μm, it is necessary to use a current density in the course of galvanic deposition of 1.3 A / mm² at an exposure of 5 hours. The subsequent increase in the exposure time did not lead to a significant increase in the thickness of the coating.

An experimental study of the durability of the Cu coating ET during EDM of 34CrNiMo6 steel showed that the coating burns out with an increase in the current strength above 8A (max value of table 3) (figure 4a).

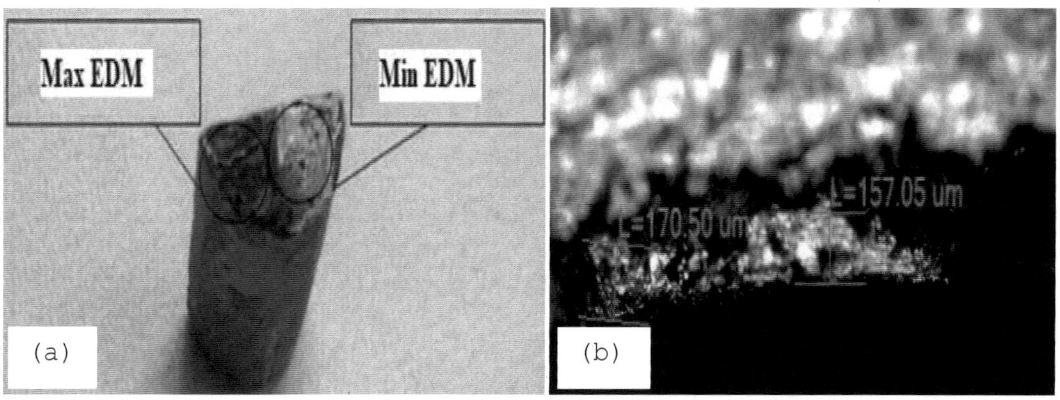


**Figure 4**. ET sample after EDM in a) max and min modes, b) the thickness of the coating after 2 hours of EDM at min mode

It was established that during EDM of workpiece with a current of 2A (min value), the coating burned out by 55% after 2 hours of processing (figure 2b). Complete burnout of the coating was witnessed after 5 hours of EDM.

On the basis of the experimental data, tests sample of cortical ET (figure 5) was fabricated and sent for production. This ET is intended for the manufacture of "Hook" type special engineering products.

**Figure 5**. As fabricated Cortical ET

The resulting electrode was fully compliant with the EDM requirements but was prone to burning of the coating surface when used in rough machining conditions. Research on accuracy of processing would be carried out in the future.

## 4. Conclusions

The process of galvanic deposition of copper on the electrode surface has been carried out. The technology of investment casting, galvanic copper deposition and modeling has been studied. An electrode tool was fabricated to perform roughing operations on the die-sinker EDM equipment.

Based on the studies, it was found that ET coated on the aluminum alloy AK12 base allows to process 34CrNiMo6 steel, providing a given processing performance. The economic efficiency of using ET with coating was 35% in comparison with the technology for obtaining a tool from solid copper billet.

**Acknowledgement**
Financial support was provided by the Russian Ministry of Education and Science (project 11.9716.2017/8.9).

## References

[1] Bains P S, Sidhu S S, Payal H S 2016 Fabrication and machining of metal matrix composites: a review *Mater. Manuf. Processes* **31(5)** 553–73
[2] Bains P S, Singh S, Sidhu S S, Kaur S, Ablyaz T R 2018 Investigation of surface properties of Al–SiC composites in hybrid electrical discharge machining *Futuristic Composites* (Singapore: Springer) pp 181–196
[3] Sohani M S, Gaitonde V N, Siddeswarappa B, Deshpande A S 2009 Investigations into the effect of tool shapes with size factor consideration in sink electrical discharge machining (EDM) process *Int. J. Adv. Manuf. Technol.* **45** 1131–45
[4] Ablyaz T R, Simonov M Y, Schlykov E S, Muratov K R 2016 Surface analysis of bimetal after EDM machining using electrodes with different physical and mechanical properties *Res. J. Pharm. Biol. Chem. Sci.* **7(5)** 974–81
[5] Khan A A, Ali M Y, Haque M M 2009 A Study of Electrode Shape Configuration on the Performance of Die Sinking EDM. *Int. J. Mech. Mater. Eng.* **4** 19–23
[6] Ablyaz T R, Shlykov E S, Kremlev S S 2017 Copper-coated electrodes for electrical discharge machining of 38X2H2MA steel *Russ. Eng. Res.* **37(10)** 20–21

# Specifics of highly concentrated heat source influence on stone casting of technogenic and mineral raw materials

**A O Artemov, S V Naumov, M N Ignatov and M F Kartashev**

Perm National Research Polytechnic University, 29, Komsomolsky prospekt, 614990, Perm, Russian Federation

E-mail: artemoff87@mail.ru

**Abstract.** Extension of mineral and technogenic raw materials scope, in particular gabbroid and basaltoid groups of Ural region is an important issue since Perm region and nearby regions contain large reserves of both natural and technogenic (waste and screening from mining complexes) raw materials. One of possible ways to get new types of products from mineral raw materials is plasma arc treatment that is a highly concentrated heat source. Using plasma arc for the synthesis of stone casting silicate-oxide components from natural and technogenic raw materials makes producing new elements of various functional purpose possible. Varying technological parameters of the synthesis enables to produce materials of different shapes - granules, powder, fiber, and cast material.

## 1. Introduction

It is known that production process of a homogeneous oxide-silicate melt under low-temperature plasma conditions has two stages, namely, simultaneous melting of all components at 1800 - 2000 °C temperature and their homogenization that happens with temperature above 2000 ° C. At the same time melt production process at technical heating rates is also known and it consists of four stages: eutectic melt formation, dissolution of refractory components in this eutectic melt, production of heterogeneous melt and its homogenization [1, 2]. Using direct action plasma arc technology differs from the known technologies in the stage of melt homogenization that is not fully achieved because of high speed of reactions, dynamic effect of plasma gas (air) and as a result of high cooling rate as well as refractory components contained in the initial mixture.

## 2. Selection of raw materials and technical parameters of products

The silicate charge under consideration contains oxides SiO2, $TiO_2$, $Al_2O_3$, FeO + $Fe_2O_3$, MnO, MgO, CaO, K2O + $Na_2O$, $Cr_2O_3$, has more refractory components (table 1).

**Table 1.** Composition of furnace burden based on technogenic raw materials

| Type of stone casting silicate melt | Oxide content, wt. % | | | | | | | | | |
|---|---|---|---|---|---|---|---|---|---|---|
| | $SiO_2$ | $Al_2O_3$ | $Fe_2O_3$ | FeO | MgO | CaO | $Na_2O$ + $K_2O$ | $TiO_2$ | $CaF_2$ | $Cr_2O_3$ |

IOP Conf. Series: Materials Science and Engineering **510** (2019) 012002 doi:10.1088/1757-899X/510/1/012002

| | | | | | | | | | | |
|---|---|---|---|---|---|---|---|---|---|---|
| Silicate-chromite type (hornblendite, Kachkanar mining and processing plant, Sverdlovskaya region) | 43-45 | 15-16 | 3-5 | 5-8 | 8-9 | 9-17 | 2.5-4 | 1-1.5 | 1.5-2 | 2-2.5 |

Refractory chromium oxide interacts with carbon and forms a small amount of free chromium and carbon dioxide; as chromium interacts with oxygen almost immediately (there are quite a lot of free oxygen ions, since they are released during dissociation of silicon dioxide and plasma-forming gas - air) or immediately reacts with carbon to form chromium carbide.

Chromium carbide reacts with oxygen to form chromium oxide again (also secondary) and carbon oxide and then secondary chromium oxide interacts with iron oxide (that presents in the melt as a decomposition product of magnetite), resulting in a mineral-like compound - chromite (1). Presence of such reaction is possible under conditions of high temperatures and a sufficient concentration of carbon dioxide, for example, when a conductive graphite substrate is used. Any electrically conductive material with a high melting point (copper, tungsten, graphite) can be used as a conductive substrate necessary for formation and stable combustion of a direct-action plasma arc.

In the greater mass of resulting melt there is a high concentration of oxygen ions, it is significantly higher than carbon dioxide concentration when using a graphite substrate. Calcium oxide is also present in the melt and therefore there is a reaction where chromatite is formed (2):

$$Cr_2O_3(solid) + FeO(liq.) \rightarrow FeCr_2O_4(solid) \quad (1)$$

$$2 \cdot Cr_2O_3(solid) + 4 \cdot CaO(liq.) + 3 \cdot O_2\uparrow \rightarrow 4 \cdot CaCrO_4 \quad (2)$$

$$CaO(liq.) + SiO_2(liq.) \rightarrow CaSiO_3(solid) \quad (3)$$

Viscosity of the rock melt is one of the important factors of production under plasma arc influence on purpose to produce granules or fibers. According to the results of calculations (figure 1, a), graphical dependences were obtained describing viscosity values of gabbro-diabase melts of Ural region (figure1, b) [3].

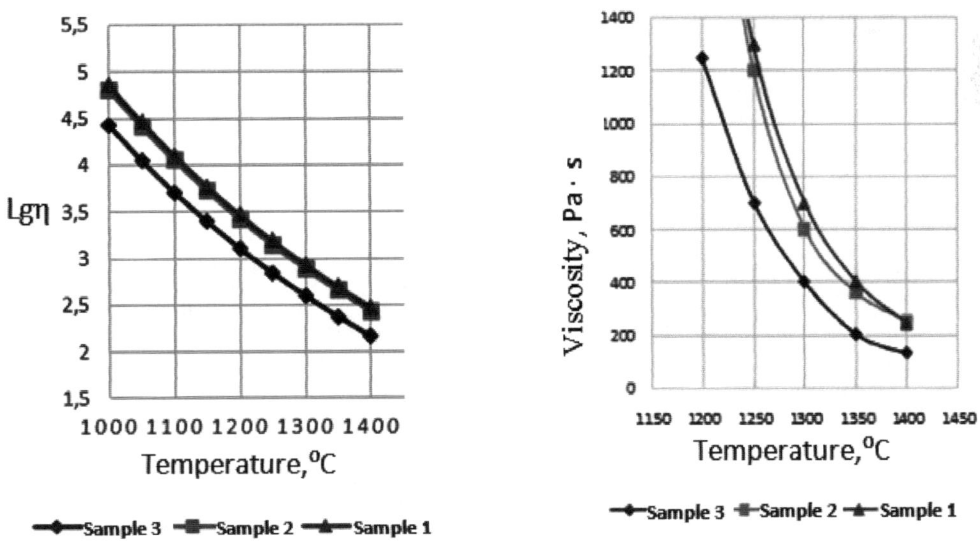


a b

**Figure 1.** Viscosity - temperature value dependency of Ural region gabbroid group raw materials: a – viscosity logarithm, lgη; b – viscosity value, Pa · s

Refractory components such as $MgFeCr_2O_4$ that harden at 1650-1800 $^0C$ as well as complex oxide compounds based on $Cr_2O_3$, CaO and $SiO_2$ (3), do not have time to melt under short plasma arc influence but reach the necessary viscosity state for plasma jet blowing forming fibers.

According to findings, viscosity of the resulting melt from mineral raw materials is 130–250 Pa · s at 1400 °C, that goes in accordance with requirements of petrurgical raw materials for mineral fiber production. However, taking into account dynamic effect of plasma arc as well as effect of plasma gas (air) that are conducive to formation of fibers, the range of the required viscosity can be wider.

## 3. Description of the experiments

Such products as mineral wool used in producing of basalt fibers, is widely known as raw material with high availability, relatively low cost, an average chemical composition, favorable melting and toughness for producing this class of products [4]. In this work, rocks of gabbroid group are used for mineral fibers production [5], that are less common due to their high melting temperatures, lower meltability and melt viscosity. However, there is a large number of screenings of these rocks available on Ural region territory and besides under high-temperature dynamic effects of plasma arc, these disadvantages can be eliminated.

Plasma technology for mineral (ceramic) fibers production is based on using a plasmatron to get a stream of high - temperature gas (air, argon, nitrogen, etc.). Furnace burden melts in a special chamber under plasma influence, Liquid melt is blown up by the gas flow and a large amount of fine mineral fibers is formed [6-8].

There are known methods [9] for producing mineral wool using gas - flame burners, however, plasma equipment has several advantages, in particular:

1) Plasmatron gives a significantly higher temperature and provides more efficient melt blown and as a result, thinner fibers.

2) Plasmatron enables temperature adjusting of in a wide range and so it is possible to select optimal modes for producing fiber.

3) Plasma arc technology can significantly reduce energy consumption.

4) Long service life of plasmatrons before replacing the electrodes (500 to thousands of hours) with a short time to replace the electrodes that is up to 15 minutes [10].

Plasma cutting system with a direct-action plasmatron was used to get laboratory samples of mineral fibers based on gabbro - rocks from Pervouralsk deposit, the arc was ignited on tungsten substrate.

For feeding plasma jet to the zone of action gabbro-rocks were crushed to the state of a finely dispersed charge of fractional composition up to 0.1 mm and mixed with liquid sodium glass so that the charge did not scatter under plasma jet influence and also melted easier and formed fibers.

Figure 2 (a) shows how a particle is lengthened under the action of high temperatures and plasma jet dynamic effect, forming a fiber [11].

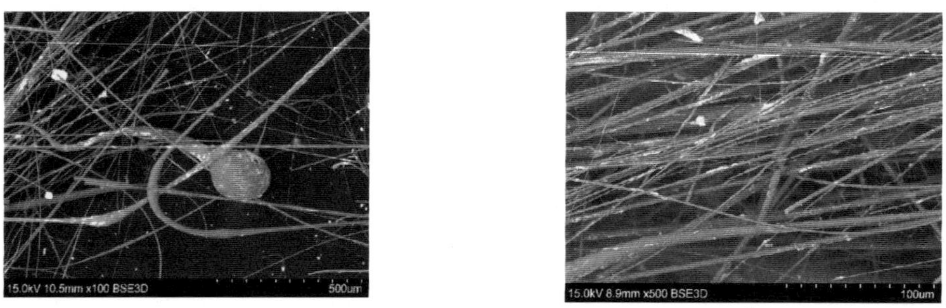


a b

c d

**Figure 2.** Gabbro-rocks mineral fiber (hornblendite): a – granule that is not lengthened and blown up to fiber under plasma jet influence, x100, b – lengthened fiber in one direction, x500, c – fiber, x3000, d – fiber with deposited oxide layer of tungsten x10000.

A distinctive feature of production of fibers is melting of charge in a direct-action plasma arc. Despite the fact that plasmatron is designed for air plasma cutting of metals, there is possible easy adjustment of equipment for mineral fiber production. The thickness of fibers is in a wide range from 500 nm to 20 μm (Figure 2 b, c). The thickness of fibers is affected by many parameters, including feeding rate of the charge on the binder, the ratio between gabbro-rocks and sodium liquid glass, fractional composition of the charge, flow rate and gas pressure, the mode of operation of the plasmatron, etc.

There is a deposited finely dispersed layer with particles less than 100 nm on the surface of the fibers. Scanning electron microscopy with microprobe analysis showed that chemical composition of the layer consists mainly of tungsten oxide (W = 70.6 wt.%). Particles found on the fiber in form of solid components of aerosols are formed as a result of a partial burning out of tungsten substrate that the arc was ignited on. Oxidation of metallic tungsten in oxygen atmosphere at temperature above 500 ° C [12, 13] goes by the reaction

$$2 \cdot W + 3 \cdot O_2 \rightarrow 2 \cdot WO_3 \quad (4)$$

Formation of tungsten oxide on the surface of fibers is confirmed by presence of yellow pigment on fibers and it is often used for coloring glasses and ceramics that are adjacent in composition and structure of products.

Using granulation pool and adjusting thermal regime of plasma arc (power, speed and pressure of gas supply), granulated welding flux can be obtained from the natural and technogenic raw materials of gabbroid group in a wide range of fractional composition.

After plasma arc influence, produced welding flux granules are homogeneous and have a uniform favorable spherical shape with particle size from 0.2 to 3 mm that corresponds to fractional composition of welding fluxes [14].

In granulation process of mineral raw materials based on gabbroid group there is a slight change in chemical composition of granules, but the processes leading to the change are well studied and can be predicted with the further development of welding materials with a specified chemical composition.

Silicate-oxide materials are in demand in many industries, including defense industry, however, their use is limited by forming technological possibilities. Therefore, the expansion of range of technological solutions for shaping products from silicate-oxide materials is an important issue [15].

Using plasma arc to melt furnace burden of natural or technogenic raw materials can significantly speed up the melting process with less energy (on average in 2 times). Silicate-oxide melt is formed under conditions of constant influence of plasma jet to the charge in the melting pot, that can be shaped into a product by casting method.

Stone casting silicate-oxide material produced in this way has high physical and mechanical characteristics (strength, hardness, wear resistance, dissipativity) compared with ceramic materials of similar composition.

These properties are achieved by high plasma temperature influence (10.000 - 15.000 K), which increases degree of homogeneity of the melt, eliminates the formation of porosity, and contributes to favorable factors for forming a fine-grained structure.

## 4. Conclusion

There is a proved possibility of using plasma arc for the synthesis of stone casting silicate-oxide materials from raw materials of gabbroid rocks of Ural region, both natural (gabbro diabase) and technogenic origin (hornblendite), that enables producing new materials of various functional purpose. Despite the fact that it is difficult to get silicate-oxide melt from this raw material due to technological features, for example, presence of sodium oxide and potassium in the form of veins that impede raw material stable melting, increased melting temperature and lower viscosity in comparison with basalt, using plasma technologies in producing stone casting silicate-oxide materials from gabbro-rocks solves the problems outlined above. By varying technological parameters of the synthesis, it is possible to obtain materials of different shapes - granules, powder, fiber, and cast material. Size of produced fibers and granules depend on many factors, but the most important are physical and chemical properties of raw materials and technical characteristics of the plasmatron.

## Acknowledgments

The reported study was funded by RFBR according to the research project № 18-33-00599.

## References

[1] Iuriev I Y et al. 2015 Physical chemical studies of dispersed aluminosilicate wastes for obtaining the burned building materials *IOP Conf. Series: Materials Science and Engineering* **71**
[2] Volokitin O G, Volkitin G G and Shekhovcov V V 2015 Physicochemical Processes at Plasma Treatment of Quartz-Feldspar Raw Materials *Advanced materials research* **1085** 474–77
[3] Persikov A S 1984 *Viscosity of igneous melts* (Moscow: Nauka) p 504
[4] Buyantuev S L, Kondratenko A S et al. 2016 The study of the mineral and chemical features of the trans-baikal basalt deposit to generate melt using the electric arc method *Biosciences Biotechnology Research Asia* **13** (**2**) 1063-68
[5] Artemov A O, Ignatov M N, Ignatova A M and Naumov S V 2017 Composition development and production technology of stone casting silicate materials and items *Key Engineering Materials* **743** 401-05
[6] Denisov G A et al. 2003 *RU Patent 2214371* C2
[7] Popowski V M et al. 2003 *RU Patent 2211193* C2
[8] Volokytin O G et al. 2009 *RU Patent 2355651* C2
[9] Niehoff T et al. 2013 *RU Patent 2498949* C2
[10] Information on http://www.forec.ru/plaz/wool-eng.htm
[11] Jain N, Singh V K and Chauhan S 2017 Review on effect of chemical, thermal, additive treatment on mechanical properties of basalt fiber and their composites *Journal of the Mechanical Behavior of Materials* **26** (**5-6**) 205-11
[12] Lassner Erik and Wolf-Dieter Schubert 1999 *Tungsten: Properties, Chemistry, Technology of the Element, Alloys, and Chemical Compounds* (New York: Kluwer Academic)
[13] Knunyants I L 1988 *Tungsten oxides, Chemical Encyclopedia* **1** (Moscow: Soviet Encyclopedia)
[14] GOST 9087-81. *Welding fluxes. Technical specifications*
[15] Artemov A O, Ignatov M N and Ignatova A M 2017 Influence of structural components on strength properties of silicate stone casting materials during controlled crystallization *Solid State Phinonema* **265 SS** 1148-51

IOP Conf. Series: Materials Science and Engineering **510** (2019) 012003 doi:10.1088/1757-899X/510/1/012003

# Resource-efficient modular space greenhouse with LED lighting

**D V Bylkov and A D Kapralov**

Division for Material Science, School of Advanced Manufacturing Technologies, National Research Tomsk Polytechnic University, 30 Lenin Avenue, Tomsk 634050, Russia

Email: prototype103faza@gmail.com

**Abstract.** The main goal of this project is the developing a design and configuration suitable for growing plants in microgravity conditions. For a decision, a block-modular design has been developed that will maximize the efficiency of using the growing surface due to the location of the plants on all faces. The resulting cubes can be cascaded to link with each other, increasing the desired volume. As a result, the installation will use less space at the station, than the modern using analogues. Design development was made by Autodesk Inventor software.

## 1. Introduction

Nowadays, science and research programs in space develops very fast and ambitious. One of research areas is the study of the influence of space conditions (microgravity, low pressure, cosmic radiation) on plants [1-7]. Studying how plants react to life in space, we can learn more about how they adapt to environmental changes and thus understand the mechanism of functioning of plants in space, because this can be the beginning of the era of colonization and development of various planets. So, to speed up the studying of plants, a large number of experiments and research are needed, in this way, productive phytotrons are required for carrying out experiments in large volumes and at lower costs. Modern phytotrons cannot provide adequate performance, their useful area and the efficiency of growing plants does not allow making the required amount of experiments, which leads to plural deliveries of prototypes and delay in experiments. For example, delivery to orbit of a cargo weighing one kilogram costs about $ 20,000 [14]. Thus, there is a great need to develop a new phytotron layout principle.

## 2. Materials and methods

The main idea to increase the performance of the phytotron is a noticeable increase in the effective area of plant growth, in contrast to the existing options for phytotron installations for space.

Thus, the layout method was put forward in the form of a cubic form. This installation will occupy less space at the station, unlike its already existing counterparts, due to the location of plants on all faces. In other words, the entire internal surface of the module will be used, which will make it possible in theory to achieve a six fold increase in production and, accordingly, to use six times less light sources.

IOP Conf. Series: Materials Science and Engineering **510** (2019) 012003 doi:10.1088/1757-899X/510/1/012003

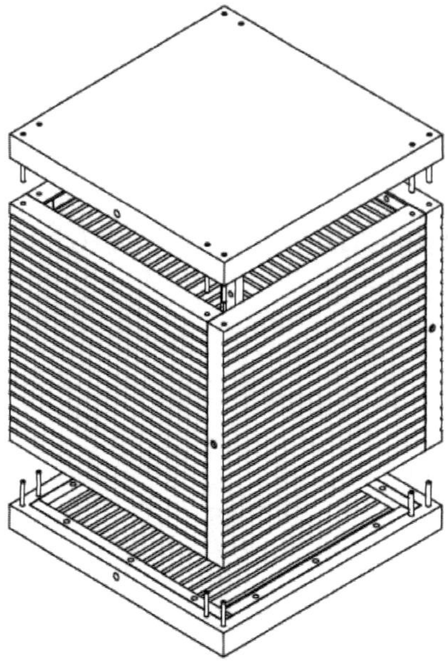

**Figure 1.** Total view of installation

The module will consist of attachable cells, which in turn will have the following components:

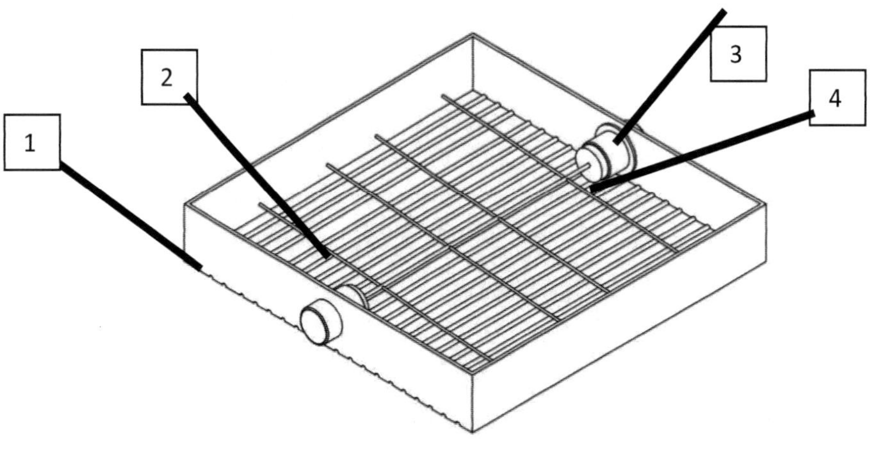


**Figure 2.** Base building sell. 1) Housing - pallet. 2) Receiver for spraying nutrient solution. 3) Pipe with seal. 4) Bung.

In assembled condition, the cell will be a sealed plate with a substrate soaked in nutrient solution inside to hold the seeds. The cell will be covered with a special sheet in order to allow the seeds to germinate inside the cube. The substrate will be needed for tight retention of the root system and proper growth of the plant, as it will not allow the liquid to spread, but will evenly deliver nutrients to the roots.

Another distinctive feature of this installation is a special arrangement of the lamp, which is made in the form of two hemispheres fastened together, located in the center of the installation. This configuration will allow you to evenly distribute the light energy and significantly reduce the cost of exposure.

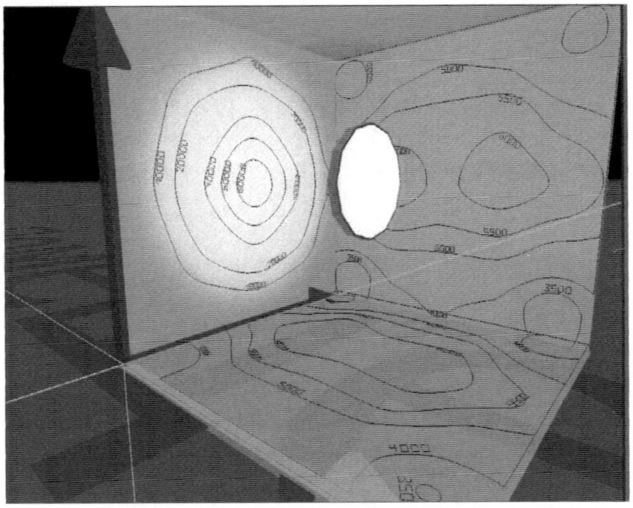

**Figure 3.** Principle of inside lightning.

Based on studies of the influence of the photon flux density on the growth rate of lettuce leaves, an irradiation regime was developed [8, 9]. The first eight days to create a photon flux density level - 80 $\mu mol \cdot s^{-1}/m^2$, and starting from day 9 until the end of the growing season of the salad - 160 $\mu mol \cdot s^{-1}/m^2$. The light source in the irradiation facility will use LEDs with a radiation wavelength within the maximum photosynthetically active radiation (400-700 nm.) [8, 11 - 13]. The structure of the light source includes a red LED with a spectral maximum of radiation at a length of 650 nm; and a white LED made on a blue crystal with an intensity maximum at a wavelength of 450 nm [10].

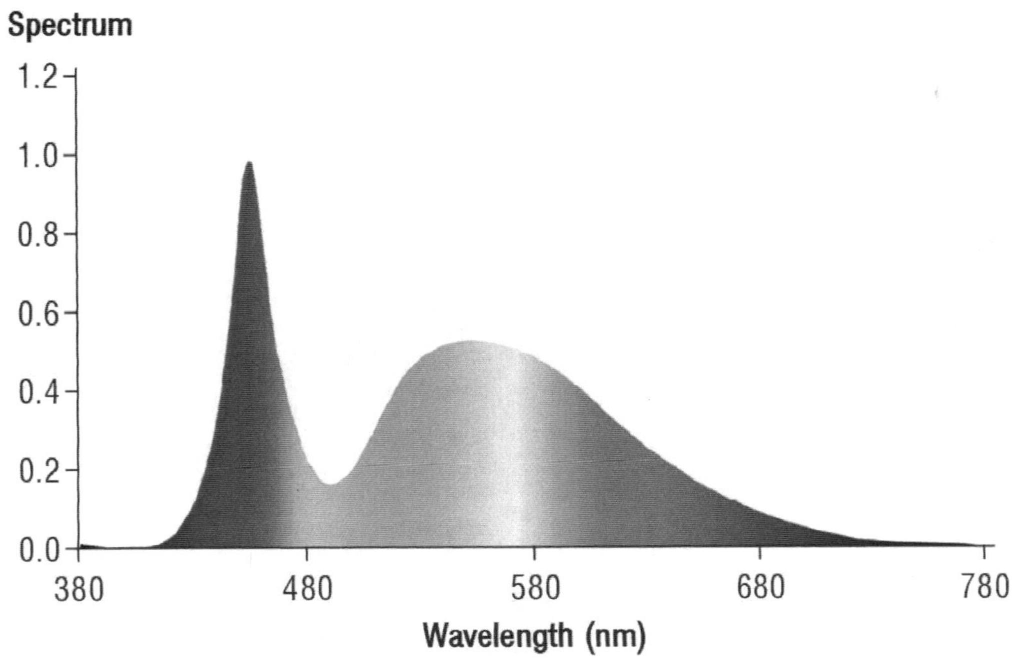


**Figure 4.** Spectrum of white LED.

IOP Conf. Series: Materials Science and Engineering **510** (2019) 012003 doi:10.1088/1757-899X/510/1/012003

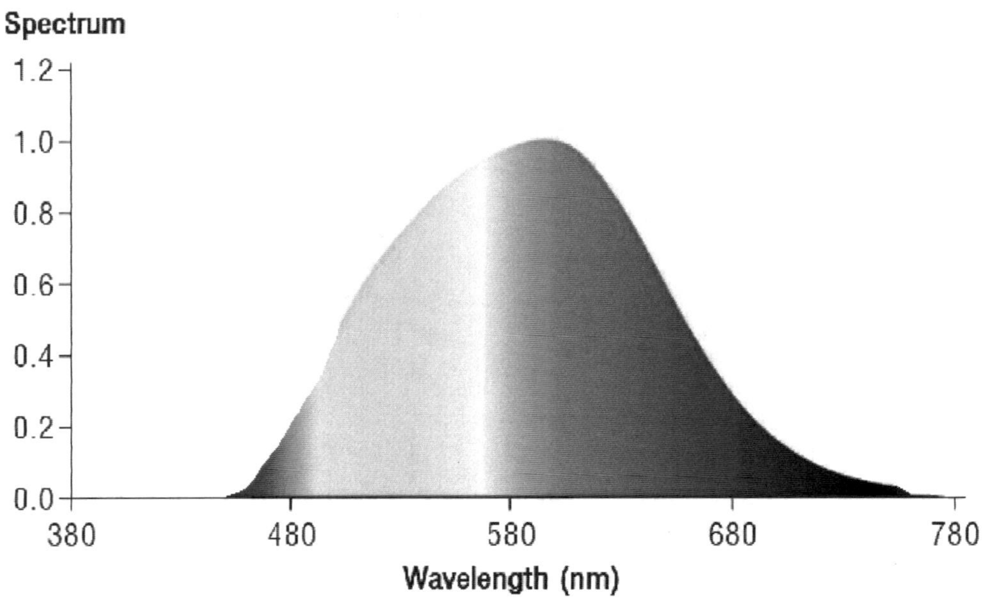


**Figure 5.** Spectrum of red LED.

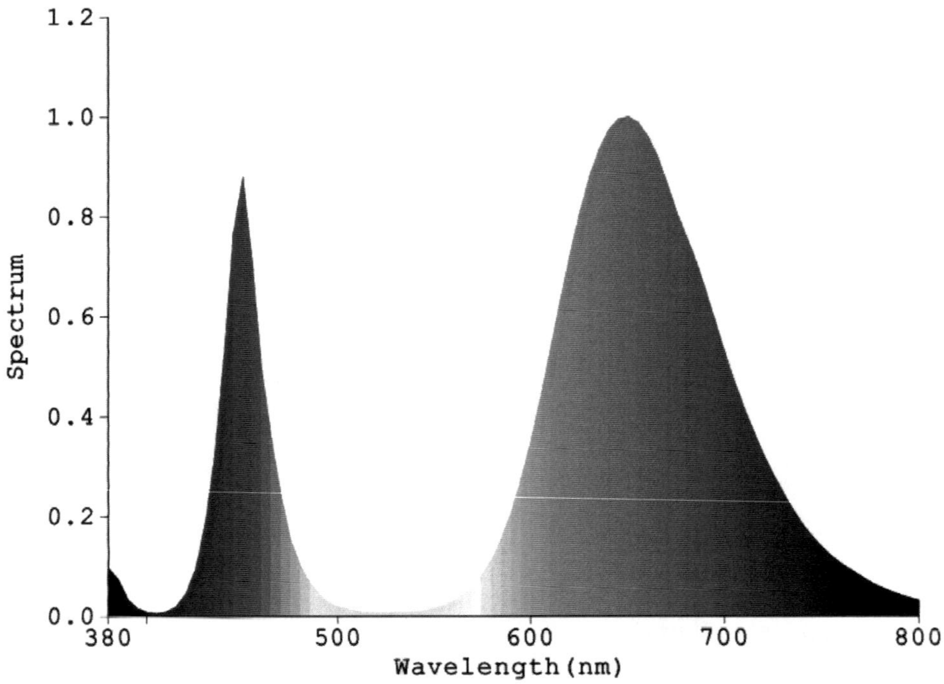


**Figure 6.** Total spectrum of lighting installation.

## 3. Conclusion

This work is a preparatory stage for the creation of a phytotron; it contains an analysis of information that makes us understand about the device of the phytotron as a whole.

Since the work is a design project, it will help us determine the form and layout of the phytotron. Of course, there is still a lot of work to be done on the development of the internal device of the phytotron, but thanks to this work, it is much easier to evaluate the ways of interaction between the parts of the phytotron, to think over the internal details, based on the overall design and construction.

So, this big step was made on the way to the development of this product, namely: the form of the phytotron was chosen, the principle and advantage of this form was explained; the main feature of which is the modularity of the design and internal use of the faces, the material and method of plant nutrition was chosen. All this allows us to qualitatively put together all further details that will ensure the functioning of the phytotron, as well as evaluate its capabilities.

**Acknowledgement**
The work was supported by the Ministry of Education and Science of the Russian Federation under the State Program in the field of scientific investigations №13.3647.2017/PP

**References**

[1] Gupta S D and Jatothu B 2013 Fundamentals and applications of light-emitting diodes (LEDs) in in vitro plant growth and morphogenesis *Plant Biotechnol.* **7** 211
[2] Bantis F, Smirnakou S, Ouzounis T, Koukounaras A, Ntagkas N and Radoglou K 2017 Current status and recent achievements in the field of horticulture with the use of light-emitting diodes (LEDs) *Scientia Horticulturae* **235** 437
[3] Singh D, Basu C, Meinhardt-Wollweber M and Roth B 2015 LEDs for energy efficient greenhouse lighting *Renewable and Sustainable Energy Reviews* **49** 139
[4] Morrow R C 2008 LED lighting in horticulture *Hort Science* **43** 1947-50
[5] Yeh N and Chung J P 2009 High-brightness LEDs – energy efficient lighting sources and their potential in indoor plant cultivation *Renew Sustain Energy* **13** 2175-80
[6] Li Q and Kubota C 2009 Effects of supplemental light quality on growth and phyto- chemicals of baby leaf lettuce *Environmental and Experimental Botany* **67** 59
[7] Samuolien G, Brazaityt A, Jankauskiene J, Vir A, Sirtautas R, Novickovas A, Sakalauskiene S, Sakalauskaite J and Duchovskis P 2013 LED irradiance level affects growth and nutritional quality of Brassica microgreens *Cent. Eur. J. Biol.* **8 (12)** 1241-49
[8] Temnik S E, Smolyakova K V and Turanov S B 2016 Development of an effective quantum radiation regime for greenhouse lettuce *Intellectual energy systems: Proc. of the IV International Youth Forum, October 10-14, 2016, Tomsk* (Tomsk: TPU Publishing House) **1** pp 350-352
[9] Turanov S B, Kozyreva I N and Yakovlev A N 2014 Methods for assessing photosynthetically active radiation *Modern equipment and technologies: Proc. works of the XX international scientific and practical conference of students, graduate students and young scientists, Tomsk, April 14-18, 2014* (Tomsk: TPU Press) **1** pp149-150
[10] Bula R J, Morrow R C, Tibbits T W, Barta R W, Ignatius R W and Martin T S 1991 Light emitting diodes as a radiation source for plants *Hort Science* **26** 203
[11] Brazaityte A, Ulinskaite R, Duchovskis P, Samuoliene G, Šikšnianiene J B, Jankauskiene J, Šabajeviene G, Baranauskis K, Staniene G, Tamulaitis G, Bliznikas Z and Žukauskas A 2006 Optimization of lighting spectrum for photosynthetic system and productivity of lettuce by using light-emitting diodes *Acta Hortic* **711** 183
[12] Li H, Tang C, Xu Z, Liu X and Han X 2012 Effects of different light sources on the growth of non-heading Chinese cabbage (Brassicacampestris L.) *J. Agric. Sci.* **4** 262
[13] Korepanov V I, Omirova N I and Omarkhan A Sh 2016 LED illuminator for greenhouses (Tomsk: TPU Publishing House) pp 372-377

[14] Dolgaleva O V 2018 *Design-project of modular system of phytotron for use in home conditions: bachelor's work* (Tomsk: National Research Tomsk Polytechnic University)

IOP Conf. Series: Materials Science and Engineering **510** (2019) 012004 doi:10.1088/1757-899X/510/1/012004

# The implementation of post-processing algorithm for ultrasonic testing of welds

**I Petrov, A Vdovenko, D Dolmatov and D Sednev**
National Research Tomsk Polytechnic University, 30 Lenina Street, 634040 Tomsk, Russia

E-mail: dolmatovdo@tpu.ru

**Abstract**. The current development of systems of pulse-echo ultrasonic testing systems is conditioned by the increasing of inspection results reliability. The application of Synthetic Aperture Focusing Technique (SAFT) is one of the solutions to this problem. The implementation of the post-processing algorithms based on SAFT strongly depends on the conditions of the ultrasonic scanning. In this paper, the issues related to SAFT implementation in the case of ultrasonic inspection of welds are considered. As a result, the post-processing algorithm, which takes into account the main features of such objects inspection, is proposed. The capabilities of the suggested algorithm were verified via the computer simulations. It was determined that the suggested algorithm is able to provide accurate and precise imageries of the flaws located in the weld.

## 1. Introduction

Pulse-echo ultrasonic testing is widely applied in different fields of industry. Its advances are the high quality of testing results, safety of testing procedure and simplicity of equipment application. One of the biggest problems of this method is the complexity to estimate the real size and shape of determined flaws [1].

Non-destructive testing of welds is one of the most common ways of ultrasonic testing application. Ultrasonic testing of welds is related with the numerous of the features. The ultrasonic scanning of the weld and heat-affected zone is carried out from the surfaces of parent material. Inclined transducers or wedges made from known material are used for this purpose. In addition, a large number of indicators are present in the echo-signal as a result of the reflection of ultrasound from the bottom of parent materials [2].

Thus, the main problem today is to improve the quality of ultrasonic flaw detection for the inspection of welds. One of the most promising ways to solve this task is the application of methods and devices that visualize the internal structure of the testing object. For this purpose Synthetic Aperture Focusing Technique (SAFT) was developed. For the quality performance of this technique in the case of welds inspection, all features related to ultrasonic testing of such objects should be taken into account.

## 2. Theory

SAFT is used for improving the quality of images and implies the joint post-processing of signals received by the transducer from all its positions on the object surface. As a result, an ultrasonic aperture of large wave dimensions is synthesized and the image of the internal structure of the test object is reconstructed via the digital focusing at each point of the visualized region.

To implement this SAFT on practice, it is necessary to consider the propagation time of the ultrasonic wave in the testing object. The calculation is made according to formula (1). In the formula $c_1$ - the speed of sound; in the numerator, the path traveled by the signal to the defect ($tx$) and back ($rx$).

$$T = \frac{\sqrt{(x_{tx} - x)^2 + z^2} + \sqrt{(x_{rx} - x) + z^2}}{c_1} \quad (1)$$

During the ultrasonic testing of welds, wedges are placed between the transducer and the test object. In this regard, in SAFT implementation it is necessary to take into account the refraction of ultrasonic waves at the interfaces between two media (wedge-object) and the change of the propagation path of the wave. In the figure 1 the center of ultrasonic transducer is located in (0;0) coordinate, the coordinates of flaw is (x; y), the point of entry of the ultrasonic wave into the test object is (x1; y1), the reflection of the ultrasonic wave from the bottom occurs at the point (x2; y2).

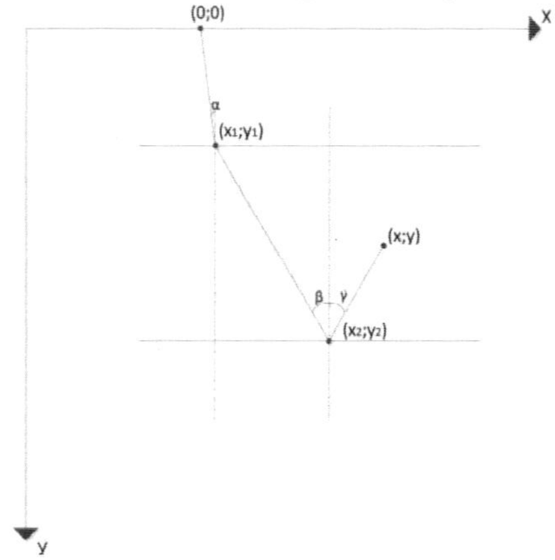


**Figure 1.** Trajectory of ultrasonic wave propagation and its reflection in testing object.

The process of propagation and refraction of waves through the interface of two media is described by the Snell's Law. The known wave velocities in the media are used for the one-way travel time calculation using equation (2), where $c_1$ is the ultrasound velocity in the contact substance; $c_2$ is the ultrasound velocity in the test object at entry to it and upon reflection from the bottom; $l_1$ is the height of the transducer position; $l_2$ is the thickness of the test object; $l_3$ - height of the defect from the bottom; α is the angle of incidence of ultrasound on the surface of the test object; $\beta$ is the angle of refraction of ultrasound in the test object; $\gamma$ is the angle of reflection of ultrasound from the bottom of the test object:

$$T = \frac{\sqrt{(x_1 - x_0)^2 + l_1^2}}{c_1} + \frac{\sqrt{(x_1 - x_2)^2 + l_2^2}}{c_2} + \frac{\sqrt{(x_3 - x_2)^2 + l_3^2}}{c_2} \quad (2)$$

## 3. Experiment

An algorithm of spatial-temporal processing based on the SAFT principles for visualization of the internal structure of welds was implemented in the Matlab R2016b. The algorithm took into account the features of welds ultrasonic testing associated with the propagation of ultrasound through materials through the media with different acoustic properties, their multiple reflections from the boundaries of the testing object, and the necessity to scan from the parent materials of the weld and small thicknesses of the objects being inspected.

The effectiveness of the proposed algorithm was considered with respect to improving the quality of the results obtained in comparison with the space-time processing algorithm, which takes into account only the direct reflection of ultrasonic waves from discontinuities inside the test object (space-time processing using "direct" ultrasonic waves). To perform this task, such algorithm was also developed using the Matlab software.

To solve this problem, the results of computer simulations implemented in the CIVA 2016 software package were used as the initial data. The simulations considered a 16 mm thick steel sample with a welded joint in which an artificial rectangular surface-breaking flaw defect was located. In total, three series of simulations were carried out, in which the height of this defect was varied (4, 6, and 8 mm).

A phased array transducer with sixteen elements was used as an ultrasonic transducer; the distance between the centers of the neighboring elements was 0.6 mm. The central frequency of ultrasonic signals which were generated by each element of phased array was 5 MHz. Ultrasound data were sampled in Full Matrix Capture mode which implies the application of full set of all possible transmitter/receiver combinations [3]. In the framework of the simulations, the ultrasonic converter was located on a plexiglass wedge, the angle of refraction of which was 45 degrees.

## 4. Results and discussions

The simulations made it possible to obtain data for post-processing using the developed algorithm. The results of post-processing using "direct" ultrasonic waves are presented in figure 2, similar results for visualization using reflected waves are presented in figure 3.

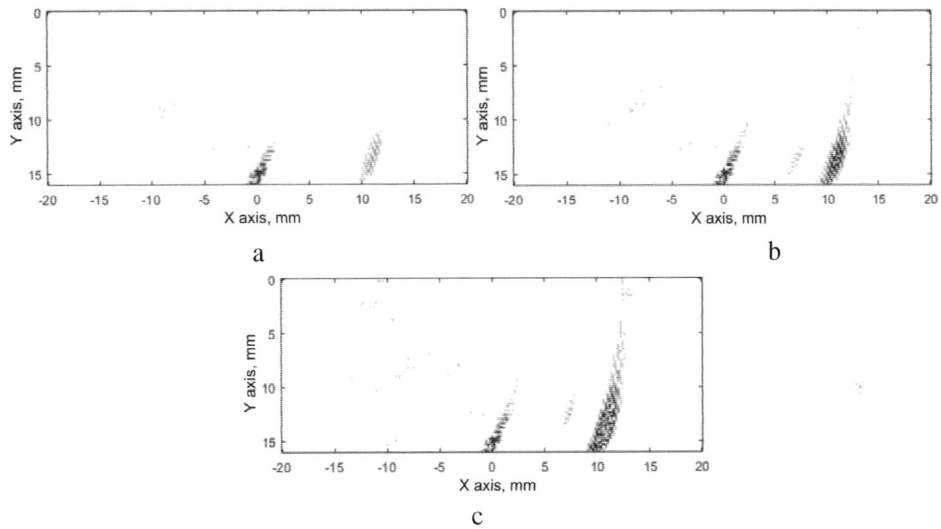


**Figure 2.** Defect image obtained using the post-processing algorithm using "direct" ultrasonic waves: a - defect height is 4 mm, b - defect height is 6 mm, c - defect height is 8 mm.

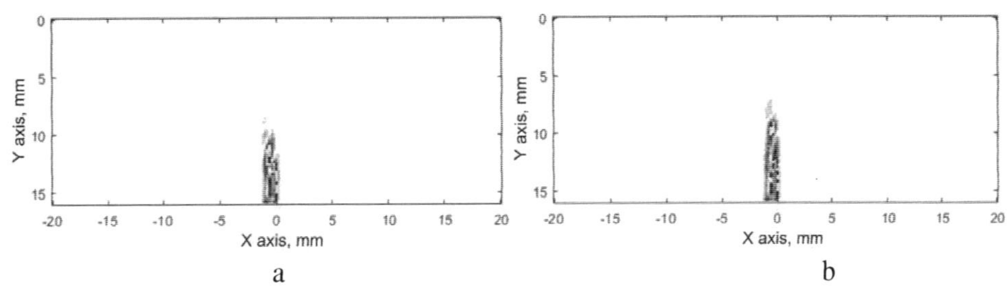

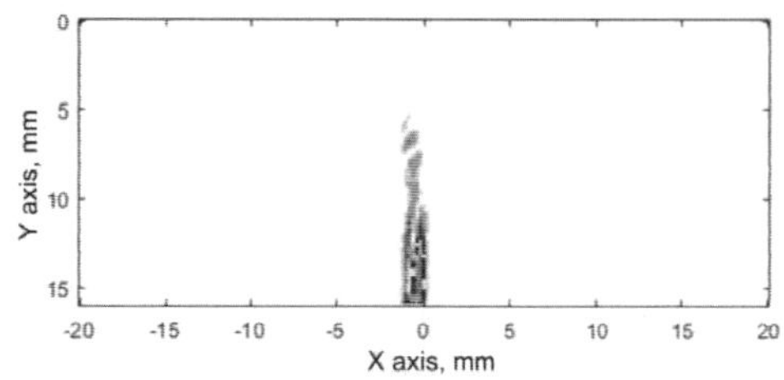


c

**Figure 3.** Defect image obtained using the post-processing algorithm using multiple reflected ultrasonic waves: a - defect height is 4 mm, b - defect height is 6 mm, c - defect height is 8 mm.

In the images of defects obtained using the post-processing algorithm using "direct" ultrasonic waves, the defect in the coordinate (0; 0) (the real position of the defect) has an undefined shape, and in the coordinate (10; 0) a false image of the defect appears. As you can see, with increasing size of the flaw, the false image becomes more intense and drowns out the real defect. The defect images obtained using the post- processing algorithm using repeatedly reflected ultrasonic waves have a clearly defined rectangular shape and equally clearly define both small (4 mm) and large (8 mm) defects.

## 5. Conclusion

The complexity of the ultrasonic testing of welds is connected with the difficulties in determining the actual size of the flaws. In this manuscript, the feasibility and effectiveness of the reconstruction were investigated, taking into account the rays reflected from the bottom of the testing object. Such an approach can be more beneficial in comparison with common SAFT algorithm implementation, which implies the post-processing of the ultrasonic waves directly retracted from the controlled object. It is related to a large number of reflections of ultrasonic waves from the boundaries of the test object, which can cause the appearing on the synthesized image false indications that do not correspond to real flaws in the testing object. The considered algorithm showed high information content and accuracy of the obtained image.

## Acknowledgements

The study was financially supported by the Ministry of Education and Science of the Russian Federation within the Scholarship of the President of the Russian Federation for young scientists and graduate students who carry out advanced research and development in priority areas of modernization of the Russian economy (№ SP-1836.2018.5 and № SP-1827.2018.2).

## References

[1] Krautkrämer J and Krautkrämer H 2013 *Ultrasonic testing of materials.* (Springer Science & Business Media) p 677
[2] Barkhatov V A 2003 *RUSS J NONDESTRUCT+* **39(1)** 23-47
[3] Doctor S R, Hall T E and Reid L D 1986 *NDT INT* **3** 163-167
[4] Holmes C, Drinkwater B W and Wilcox P D 2005 *NDT&E INT* **38(8)** 701-711

IOP Conf. Series: Materials Science and Engineering **510** (2019) 012005 doi:10.1088/1757-899X/510/1/012005

# Numerical investigation of the thermal transformation of composition powder particles in the technology of selective laser melting

**V I Bogdanovich and M G Giorbelidze**
Samara National Research University, 34, Moskovskoe Highway, Samara, 443086, Russian Federation

E-mail: bogdanovich@ssau.ru; m.giorbelidze@ssau.ru

**Abstract.** The results of mathematical model development for evaluation of thermal state of spherical particles of powder composition equally treated in the technology of selective laser melting are presented. This model considers: features of laser irradiation energy transfer to the particles having smaller size but comparable with the size of radiation spot diameter; energy transfer through the sphere upper half; material melting in the solidus and liquidus range; possibility of surface material evaporation and dependence of material thermophysical parameters on temperature. This model is adapted to be used in standard finite-element software ANSYS Transient Thermal that has been applied for numerical simulation using powder of heat-resistant nickel-chromium alloy. Process regularities have been discovered, and it was shown that the range value of particles size scatter of powder composition fraction used determines principle possibility of choosing laser treatment mode to provide high quality material after melting.

## 1. Introduction

Today new technological state of modern industry is realized due to both powerful computer-aided design systems and new technologies. Additive technology of selective laser melting (SLM) of powder materials is among them.

One of the main issues of SLM technology is reasonable choice of laser treatment mode of powder composition used. Significant amount of papers dealt with this issue [1-24], and they can be divided into two groups: papers based on experimental research results [7, 8, 10-14, 16-17] and papers based on mathematical simulations [1-6, 9, 15].

In the papers dealing with mathematical simulation, we can distinguish two levels of process – macro-scale and micro-scale. At the macro-scale level, powder layer is simulated as a solid material with effective values of physical-mechanical characteristics. The tasks of prediction of its heating and melting, thermal and residual stresses, material shrinkage and layer dimension variations are solved. At the micro-scale level, material is considered as a powder layer having various thickness and various arrangement of particles. The tasks of determining of thermal particles state, material spreading, estimation of penetration of melt into the lower layers, initiation and extinction of material voids are solved.

In papers dealing with mathematical simulation, researchers come to conclusion that the mode range of laser treatment is very narrow, and it requires considerable numerical and experimental researches to get the material of required quality.

Analysis of papers shows that potential material characteristics produced from powder under its layer-by-layer treatment are defined by the treatment mode of each separate particle of this layer. Further layer treatment by multidirectional travelling of the laser beam, thermal treatment or impregnation allows, if possible, enhancing these characteristics and leveling some negative consequences of local laser treatment of the powder layer.

The treatment mode of each separate particle is to provide at least its mandatory melting at the points of contact with other particles and the underlayer to provide topochemical interaction reactions of materials at these contact surfaces. However, under thermal flow density used and insertion of radiation through the upper half of the surface, particles of different sizes are heated up to various temperatures with substantial gradient throughout the height of their sections. This can result in heating of particles of other sizes less than melting point in the lower part and heating up to the vaporization temperature at the upper surface, but the heating temperature is optimum, if the particles are of the same size. For example, this different heating of particles of various sizes treated at the same mode allows us to conclude on insufficient informativeness of just experimental results and suggest a hypothesis on necessary use of particles just with certain scatter range of its sizes.

As applied to the tasks of mathematical simulation of thermal processes in powder composition particles, treated at the same mode of SLM technology, it is required to take into account surface evaporation, dissymmetry of laser irradiation through the particle upper surface and dependence of thermal flow absorption on its incidence angle.

Purpose of this paper is to report research results got during development of proper mathematical model of evaluation of thermal state of particles of powder composition treated at the same mode in the technology of selective laser melting and adapted to be used in the standard software ANSYS Transient Thermal. This mathematical model suits for applied use to determine optimum mode of composition treatment. Some results of numerical research of powder composition of heat-resistant nickel-chromium alloy ВВ751П are given.

## 2. Process conceptual model

In the typical SLM process, powder is put on the platform surface to manufacture a workpiece or it is put on the previously alloyed layer with successive vertical shift of the movable platform part by 30…100 μm. Powder is levelled by height and consolidated with the scraper travel that is pressed to the fixed part of the platform surface. Powder layer is treated by the laser energy beam continuously travelling the desired path along the model section contour in order to fuse the powder at this contour. Then powder fusion within the distinguished contour is carried out when the beam is traveling along the desired continuous path. At the completion stage, the layer is blown by a laminar gas flow to remove air-floated particles.

## 3. Mathematical modeling of particle heating in selective laser melting

Statement of mathematical model of the processes of heating, melting, the following crystallizing and cooling-down of the spherical particle of the metal alloy with the temperature range of the phase transition is to include corresponding differential heat conduction equation, initial and boundary conditions. Boundary condition is to consider features of heat transfer to the particle from laser irradiation and features of heat exchange between the particle and the environment.

It can be shown that typical differential heat conduction equation for an isotropic body with emitting and absorbing heat in the volume [19] can be brought to the following form considering phase material transformation at the solidus and liquidus range, possibility of surface evaporation and adapted to end-to-end finite-element procedure to be used in the standard software ANSYS:

$$c_{ef}(T)\rho(T)\frac{\partial T}{\partial t} = div\left[\lambda(T)grad(T)\right] \tag{1}$$

The following designations are used in (1):

$$c_{ef}(T) = c(T) + \frac{L}{T_L - T_S}\varphi\left(\frac{T - T_S}{T_L - T_S}\right)\left[\eta\left(\frac{T - T_S}{T_S}\right) - \eta\left(\frac{T - T_L}{T_L}\right)\right] + \frac{L_{ev}}{T_{ev}}\delta\left(\frac{T - T_{ev}}{T_{ev}}\right), \quad (2)$$

where $T(t,\vec{r})$ is temperature at the time moment $t$ of the body point having coordinate $\vec{r}$; $c(T)$, $\rho(T)$ and $\lambda(T)$ is specific thermal capacity, density and material thermal conductivity coefficient; $T_S$, $T_L$ and $T_{ev}$ is solidus and liquidus temperature and material evaporation temperature; $L$ and $L_{ev}$ is specific fusion and evaporation heat; $div$ and $grad$ is differential operators of divergence and gradient; $\eta(u)$ is unit Heaviside function ($\eta(u)=0$ when $u<0$ and $\eta(u)=1$ when $u \geq 0$) and $\delta(u)$ is Dirac delta function with corresponding dimensionless arguments; $\varphi\left[(T-T_S)/(T_L-T)_S\right]$ is rate function of phase transition (in the instant melting case it is called by analogy with the known function called crystallization rate [20]) connected with specific volume share of new liquid phase in the nominal volume of melted material in the liquid crystal area of the phase transition $\psi\left[(T-T_S)/(T_L-T)_S\right]$ by the relation $\varphi(u) = d\psi(u)/du$.

This differential heat conduction equation (1) retains the same in different body parts with and without phase transitions as well as for various stages of its heating and cooling.

Initial condition is the equality of the temperature to the initial temperature of the sphere material at the time moment $t = 0$:

$$T(0,\vec{r}) = T_0(\vec{r}). \quad (3)$$

Boundary condition is determined by the fact that laser irradiation striking the upper sphere surface is converted into the thermal flow inserting the sphere surface normally to the upper half sphere material. It acts during the period of passing over the particle. Boundary condition is also determined by convection cooling and thermal radiation of the whole surface. In a spherical coordinate system that is convenient in case of temperature distribution in the spherical particle, boundary condition is the following:

$$\lambda(T)\frac{\partial T}{\partial r} = q(t,\theta,\varphi)\left[\eta(\theta) - \eta(\theta - 0.5\pi)\right] - \alpha\left[T - T_c\right] - \varepsilon\sigma\left[T^4 - T_c^4\right], \text{ when } r = R. \quad (4)$$

The following designations are used in (4): $R$ is spherical particle radius; $\alpha$ is coefficient of convective heat exchange; $\varepsilon$ is integral radiation coefficient of the sphere surface; $\sigma$ is Stefan-Boltzmann constant; $T_c$ is gas temperature in the chamber; $\theta$ and $\varphi$ is zenith and azimuth angles of the spherical coordinate system connected with sphere center; difference of unit Heaviside functions taking into account the fact that thermal flow to the sphere surface does not equal zero only in the zenith angle area $0 \leq \theta \leq 0.5\pi$ is given in square brackets.

Thermal flow value $q(t,\theta,\varphi)$ in general case is to depend on time $t$, zenith $\theta$ and azimuth angle $\varphi$ of spherical coordinate system as laser irradiation source travels over the sphere with constant speed $v$. In addition, thermal flow density is also to depend on zenith $\theta$ as material absorption coefficient depends on incidence angle of laser irradiation. Therefore, temperature in the spherically symmetrical ball depends on the position of laser spot over the sphere, and it is not symmetrical of zenith and azimuth angle.

In this formulation boundary problem (1) – (4), if $q(t,\theta,\varphi)$, it can be solved with numerical simulation using standard software ANSYS. Results of this numerical simulation have shown that calculations are very computer time-consuming, but the result is complex to analyze and insufficiently informative to develop technological suggestions on choice of optimum treatment modes of powder composition with sizes scatter range.

IOP Conf. Series: Materials Science and Engineering **510** (2019) 012005 doi:10.1088/1757-899X/510/1/012005

## 4. Results and discussion

Numerical simulation of particle heating process introduced as a boundary problem (1-3) has been carried out using standard software ANSYS. Heating mode of the powder composition has been at heat flux $q_0$ at source traveling speed $v$. Adequacy of the mathematical model has been verified and confirmed by means of model parameters of the material based on accuracy of numerical calculations in comparison with results of known exact and approximate analytical solutions. It is also based on estimation of stability and convergence of solutions under decreasing of partition interval sizes, on calculation error of relation (4), and on comparison with the results of special experimental research.

Composition treatment mode has been chosen by the following way. According to the particle heating simulation with particles of maximum diameter in the composition (for example, maximum diameter 50 or 40 μm) when the temperature in the sphere bottom point reaches material liquidus temperature ($T_L = 1346$ °C), and the temperature in the sphere top point does not exceed material evaporation temperature ($T_{ev} = 2913$ °C) heating time of the particle has been chosen. According to this time, value of source travel speed has been determined, and heating period and heat flux that are to be applied to the particles of other diameters when treated at the same mode.

Typical results of mathematical simulation are shown in Figure 1 and Figure 2.

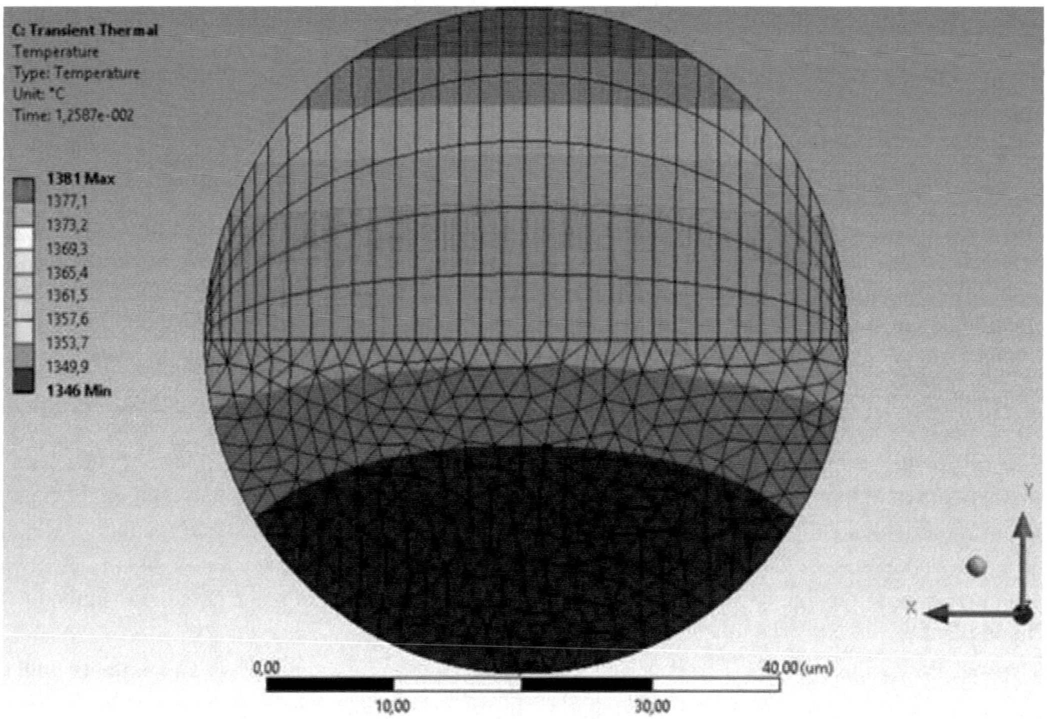


**Figure 1.** Temperature distribution of the sphere cross-section with diameter 50 μm when heat flux $q_0 = 1 \cdot 10^8$ W/m$^2$ at the time moment 0.012587 s when its temperature at the bottom point has reached liquidus temperature 1346 °C.

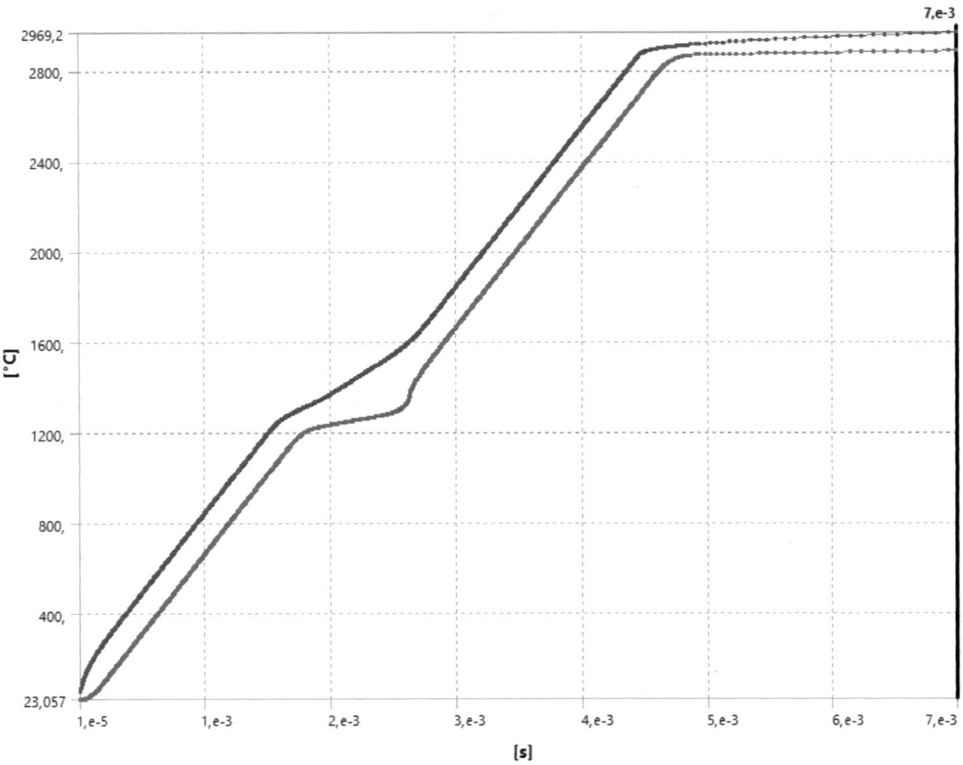


**Figure 2.** Temperature of the top and bottom points of the sphere with diameter 50 μm depending on the time period at heat flux $q_0 = 1 \cdot 10^8$ W/m$^2$.

## 5. Conclusion

It has been found out that for the powder composition with particle diameters 50…20 μm there is no treatment mode at which the particles with maximum diameter 50 μm are completely melted, and the particles with minimum diameter 20 μm do not reach boiling at intensive evaporation. As a result, some research has been carried out to choose treatment mode of powder composition with narrower particle scatter range within 40…20 μm that has shown availability of the mode with complete melting of all the particles and just since evaporation of the particles with 20 μm diameter.

As a result of numerical simulation SLM, basic treatment mode of powder composition 40…20 μm has been recommended. It provides heat flux $1.206 \cdot 10^9$ W/m$^2$ and travel speed of the laser spot 328 mm/s. Preliminary sizing of powder in delivery condition with extraction of recommended fraction is suggested.

## Acknowledgments

The reported study was funded by the Ministry of Education and Science of the Russian Federation according to research project No.11.2744.2017.

## References

[1] Ankudinov V E, Aflyatunova D D, Krivilev M D and Gordeev G A 2014 *Computer simulation of transport and deformation processes in continuous environment* (Izhevsk: Udmurtskiy universitet Publisher)

[2] Kalitkin N N 2011 *Numerical methods* (Saint-Petersburg: BHV-Petersburg Publisher)

[3] Koldoba A V, Poveshenko Yu A and Popov M V 2009 Mathematical modeling of laser sintering of two-components powder mix *Electronic journal of IPM Preprint im. M.V. Keldysh* prep2009_38
[4] Krivilev M D, Krivilev E V, Harangevskiy E V, Gordeev G A and Ankudinov V E 2010 *IPU RAN Publisher* 299-322
[5] Harangevskiy E V and Krivilev M D 2011 *Laser physics, laser technologies and methods of mathematical modeling of laser impact on matter* (Izhevsk: Udmurtskiy universitet Publisher)
[6] Shishkovskiy I V 2005 *Selective laser sintering and synthesis of functional structures* (Phd dissertation. Samara State University)
[7] Shishkovskiy I V 2015 *The foundation of high-resolution additive technologies* (Saint-Petersburg: Piter Publisher)
[8] Arisoy Y M, Criales L E, Özel T, Lane B, Moylan S and Donmez A 2017 *Int. J. Adv. Manuf. Technology* **90** 1393-1417
[9] Boley C D, Khairallah S A and Rubenchik A M 2015 *Applied Optics* **54** 2477-2482
[10] Sufiyarov V Sh, Popovich A A, Borisov E V and Polozov I A 2015 *Non-ferrous Metals* **1** 79-84
[11] Gong H 2013 *Generation and detection of defects in metallic parts fabricated by selective laser melting and electron beam melting and their effects on mechanical properties* (Phd dissertation. University of Louisville)
[12] Grasso M and Colosimo B M 2017 *Measurement Science and Technology* **28** 1-25
[13] Smelov V G, Sotov A V and Agapovichev A V 2016 *Ferrous metals* **9** 61-65
[14] Huang Q, Hu N, Yang X, Zhang R and Feng Q 2016 *Frontiers Mater. Sci.* **10(4)** 428-431
[15] Knol M F 2016 *Thermal modelling of selective laser melting* (PhD Thesis. Delft University of Technology)
[16] Popovich V A, Borisov E V, Popovich A A, Sufiiarov V Sh, Masaylo D V and Alzina L 2017 *Materials & Design* **114** 441-449
[17] Zeng K 2015 *Optimization of support structures for selective laser melting* (PhD dissertation. Louisville)
[18] Samarskiy A A and Vabishevich P N 2003 *Computational heat transfer* (Moscow: Editorial URSS Publisher)
[19] Kartashev E M 2001 *Analytical methods in the theory of thermal conductivity of solid bodies* (Moscow: Vysshaya shkola Publisher)
[20] Bochvar A A 1956 *Metallurgy* (Moscow: Metallurgizdat Publisher)
[21] Bogdanovich V I, Giorbelidze M G, Sotov A V, Pronichev N D, Smelov V G and Agapovichev A V 2017 *Proceedings of Samara scientific center of the Russian Academy of Sciences* **19(4)** 105-114
[22] Kovalev O B and Fomin V M 2013 *Physical foundation of laser cutting of thick sheet materials* (Moscow: Fizmatlit Publisher)
[23] Grigoryantz A G 1989 *Foundation of materials laser processing* (Moscow: Mashinostroenie Publisher)
[24] Sotov A V, Pronichev N D, Smelov V G, Bogdanovich V I, Giorbelidze M G and Agapovichev A V 2017 *Proceedings of Samara scientific center of the Russian Academy of Sciences* **19(4)** 96-104

IOP Conf. Series: Materials Science and Engineering **510** (2019) 012006 doi:10.1088/1757-899X/510/1/012006

# Investigation of the machinability of composite materials electrode-tools while EDM

**A O Grisharin[1*], N D Ogleznev[1], K R Muratov[1], T R Ablyaz[1], S B Preetkanwal[2] and S S Sarabjeet[2]**

[1]Perm National Research Polytechnic University, Perm-614000, Russian Federation

[2] Mechanical Engineering Department, Beant College of Engineering & Technology, Gurdaspur-143521, India

*E-mail: a.grisharin@gmail.com

**Abstract.** The aim of the work is to study the mechanical properties of composite material electrode-tools made of pseudo-alloy of the copper-colloidal graphite system for the Electrical Discharge Machining (EDM) processing of dissimilar materials. Electrode-tools made of copper, graphite and the specific composite material are compared. These electrodes were used for EDM of samples of alloyed steel and tungsten, aluminum and titanium based alloys. After processing, the surface roughness, wear and performance of the tool electrodes were evaluated. As a result of the experiment, it was established that the electrode made of a composite material, copper-colloidal graphite, in some cases, demonstrated significantly higher performance and wear resistance. Based on the data obtained, it was concluded that it is advisable to use an electrode made of a composite material to increase the efficiency of EDM processing of aluminum alloys and alloyed steels, as well as rough EDM processing of titanium alloys.

## 1. Introduction

Currently, there is an increasing trend of reducing the production time in industrial world. Manufacturing plants are quickly adapting to the release of new products, which is ensured, first of all, by a quick changeover of production. For machining of geometrically complex parts, expensive special equipments and tools are required. These manufacturing procedures require huge economic and technological investments. The use of physico-mechanical machining method is of utmost relevance, allowing the manufacture of parts with complex shapes using the three-dimensional reproduction technique [1] and EDM (Electric Discharge Machining) is the prominent choice.

The process of EDM involves the gradual melting and evaporation of conductive materials under the action of pulsating current. The molten debris, thus created is flushed off with the aid of hydrodynamic forces of the working fluid, called the 'Dielectric' [1, 2]. The ability of EDM process to machine hard-to-cut materials of any hardness is the main advantage of this technique as compared to other conventional mechanical routes, ensuring high accuracy of the resulting geometry and the surface properties of material being processed [3].

To perform EDM using the method of three-dimensional replication, die-sinker EDM machines are used, employing complex-profile electrodes as the tool. Electrolytic Copper (Grade M1 and M0) and Graphite are widely used as electrode material for EDM tool. The accuracy in the size and surface properties of processed parts using the die-sinker EDM machines depends on the geometrical accuracy of the electrode. In order to produce complex and precise geometry, the process of manufacturing an electrode could include many other mechanical and physio-technical processing operations, which makes it a time-consuming and costlier process.

The EDM technology is widely used for the manufacture of working structural elements of critical parts, such as gearbox housings, copiers, parts of hydraulic cylinders and many others. These components work under conditions of variable temperatures and dynamic loads, and therefore, high-strength materials, such as alloyed thermally hardened steels, alloys based on titanium, aluminum, tungsten and others, are used. Many of these materials have high electrical resistance, which leads to intensive wear of the electrode tools and a decrease in the performance of the process of EDM.

Considering the above, the actual task was to improve the properties, such as performance and wear resistance of electrode-tool.

## 2. Literature review

One of the effective ways to increase the wear resistance of electrode tools is the use of pseudo-alloy as composite material for their fabrication, combining their different properties [2, 3, 4]. Pseudo-alloy refers to the system of immiscible components, that is, a system in which there is a tendency to stratification in the liquid state already and there is virtually no solubility in the solid state. The microstructure of pseudo-alloys is a thin uniform mixture of two or more phases, one of which has a high reinforcement and the second has high electrical conductivity. The low-melting phase melts the refractory phase and the physical properties of the phases are additively added.

Currently, a number of wear-resistant composite materials for electrical purposes have been developed, which can significantly improve the electro-erosion properties of electrode tools as compared to the traditionally used materials. One of the most promising materials for solving this problem is the composite material comprising of copper-colloidal graphite having a graphite content of less than 2% [5]. The results of the performance and wear resistance of the electrodes are shown in figures 1 and 2.

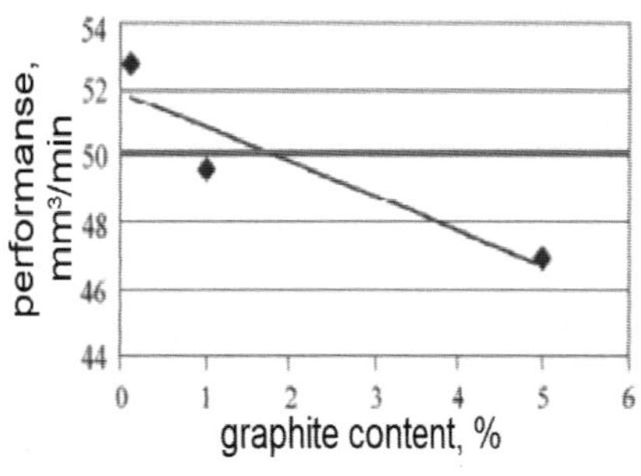


**Figure 1.** Performance of tool with graphite content

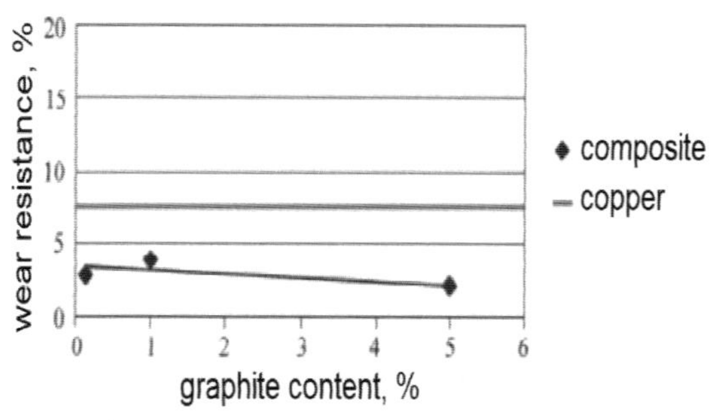


**Figure 2.** Wear resistance with graphite content

The particular pseudo-alloy is a mechanical mixture of copper and dry colloidal graphite powders, pressed and sintered by powder metallurgy technique. The pseudo-alloy components practically do not fuse together with each other in the temperature range achieved during the EDM process [6].

Production of electrode tools made of composite material by powder metallurgy method is technologically advanced technique. It can be carried out on commercially available equipments when the constituent materials are available in the form of powder with an average particle size of less than 40 microns [7]. It becomes possible to vary the chemical composition, dispersion and technological parameters of the powders used for developing the composite materials for a tool electrode by this technique, which makes it possible to significantly enhance its properties [8].

Despite the advantages of a composite material such as a pseudo-alloy system of copper-colloidal graphite, this material does not find wide application for the manufacture of complex-profile tool electrodes. This is due to the fact that the operational properties of this material during the EDM processing of various steels and alloys are not fully understood, and there are no practical recommendations.

Thus, the analysis of the literature has shown that there is a need to study the machining properties of electrodes made of composite materials for the EDM processing of dissimilar alloys.

Hence, the objective of this study was to investigate the wear resistance and performance of electrode tools made of a composite material such as a pseudo-alloy system of copper-colloid while EDM of titanium, aluminum, tungsten alloys and alloyed steel.

## 3. Materials and methods

For the experiment, electrodes were made of copper, graphite, and a composite material such as a pseudo-alloy of the copper – colloidal graphite system with a graphite content of less than 2%.

Copper powders PMS-1 GOST 49-60-75 and dry colloidal graphite C-1 grade TU 113-08-48-63-90 were used for the fabrication of an electrode tools. The powders were mixed in the required proportion in a mixer with a displaced axis of rotation for 4 hours, after which samples of 6x6x50 mm were formed from the mixture at a pressure of 600 MPa. Then, the samples were annealed in a vacuum oven at 700 °C and pressed again. After re-pressing, the samples were finally sintered in a vacuum oven at a temperature of 1000 °C for 2 hours.

For the manufacturing of an electrode tool made of composite material, copper powders PMS-1 GOST 49-60-75 and dry colloidal graphite C-1 grade TU 113-08-48-63-90 were used. After mixing the powders in the required proportion, the samples of 6x6x50 mm were developed at a pressure of 600 MPa. Afterwards, the samples were annealed in a vacuum oven at 700 °C and were re-pressed. The samples were finally sintered in a vacuum oven at a temperature of 1000 °C for duration of 2 hours.

The specimen of titanium alloy OT-4, aluminum alloy AK6, tungsten-based alloy VK8 and low alloy steel 35GS were prepared as samples of dissimilar materials.

The study of functional properties of the electrodes was carried out on a die-sinker EDM machine (Electronica Smart CNC) on rough processing modes. IPOL SEO 450 was employed as the EDM working fluid.

The performance of the electrode was estimated by the depth of the hole machined in the sample, and wear was measured by recording the difference in weight of electrode before and after machining. The surface roughness of the machined samples and electrodes was measured on a Mahr Perthometer S2 profilometer.

## 4. Results

The results of the EDM of the AK6 alloy are presented in Table 1. An analysis of the results of EDM processing of the AK6 aluminum alloy showed that the efficiency of the electrode-tool made of composite material is 1.1 times lower than that of the electrodes made of copper and graphite. When assessing the wear resistance, it was observed that the electrode-tool made of composite material possesses 3.5 times higher wear resistance than the electrode made of copper. During the experiment, the weight of the graphite electrode increased after treatment. There was adhesion of the workpiece material on the working surface of the electrode tool. Evaluation of the obtained roughness of the treated surface showed that the highest quality was provided by the electrode made of graphite.

**Table 1.** The experimental results of EDM of AK6 alloy

| | Part roughness (Ra) | Electrode roughness (Ra) | Electrode wear (gr.) | Depth of processing (mm.) |
|---|---|---|---|---|

| | | | | |
|---|---|---|---|---|
| **Cu** | 8.3131 | 0.9872 | 0.0014 | 1.2121 |
| **Gr** | 6.9828 | 0.9310 | -- | 1.24 |
| **Cu-Gr** | 8.3687 | 0.8480 | 0.0004 | 1.08 |

The experimental results of the EDM of steel 35GS are presented in Table 2. Analysis of the results while EDM of low-alloy steel 35GS showed that the electrode-tool made of composite material shows 2.4 times higher performance than the copper electrode and graphite, as well as 3.8 times higher wear resistance than the electrode from copper. Evaluation of the surface roughness showed that the electrode made of a composite material provided a lower surface quality.

**Table 2.** The experimental results of EDM of 35GS steel

| | Part roughness (Ra) | Electrode roughness (Ra) | Electrode wear (gr.) | Depth of processing (mm.) |
|---|---|---|---|---|
| **Cu** | 5.5095 | 1.0256 | 0.0023 | 0.1120 |
| **Gr** | 5.4083 | 1.4088 | -- | 0.1124 |
| **Cu-Gr** | 6.7746 | 2.8705 | 0.0006 | 0.276 |

The experimentation results of EDM of titanium alloy OT-4 presented in Table 3. Analysis of the results showed that while processing of titanium alloy OT-4, the productivity of a composite electrode tool was 21.2 times higher than that of a copper electrode and 3.9 times higher than that of a graphite electrode, but the surface roughness was enhanced significantly than other electrodes.

**Table 3.** The experimental results of EDM of titanium alloy OT-4

| | Part roughness (Ra) | Electrode roughness (Ra) | Electrode wear (gr.) | Depth of processing (mm.) |
|---|---|---|---|---|
| **Cu** | 2.8615 | 2.4104 | 0.0048 | 0.0127 |
| **Gr** | 3.9301 | 2.5033 | -- | 0.069 |
| **Cu-Gr** | 5.1555 | 1.9380 | 0.0035 | 0.2691 |

The results obtained after EDM of VK8 alloy based on tungsten are presented in Table 4. Analysis of the results of EDM processing of the VK8 alloy showed that the electrode-tool made of composite material has a significantly lower performance indicator than the copper electrode and the graphite electrode, as well as lower wear resistance compared to the copper electrode. Since the weight of electrode made of graphite was higher, it was not possible to estimate its wear. The quality of the machined surface also turned out to be significantly lower when treated with a composite electrode.

**Table 4**. The experimental results of EDM of VK8 alloy

| | Part roughness (Ra) | Electrode roughness (Ra) | Electrode wear (gr.) | Depth of processing (mm.) |
|---|---|---|---|---|
| **Cu** | 0.4517 | 2.6592 | 0.0099 | 0.0095 |
| **Gr** | 2.7619 | 2.9384 | -- | 0.0112 |
| **Cu-Gr** | 8.6498 | 6.1635 | 0.0326 | 0.0024 |

## 5. Conclusions

Based on the results of the experiment and analysis of the data obtained, it can be concluded that the use of a composite material such as a pseudo-alloy system of copper-colloidal graphite with graphite content of less than 2%, in some cases, can be an effective solution to improve the wear resistance of electrode tools and the performance of the EDM process. In particular, it is advisable to use this material for the EDM processing of aluminum alloys and alloyed steels, as well as for the rough EDM operations of titanium alloys. For the Electrical Discharge Machining of tungsten based alloys, the use of electrodes made of composite material such as a pseudo-alloy system of copper-colloidal graphite is not advisable.

**Acknowledgments**
Financial support was provided by the Russian Ministry of Education and Science (project 11.9716.2017/8.9).

**References**

[1] Bains P S, Sidhu S S, Payal H S 2016 *Fabrication and machining of metal matrix composites* (Kapurthala: IKG Punjab Technical University) p 533
[2] Smolentsev V 1983 *EDM and electrochemical methods of processing materials* vol 2 (Moscow: Higher School) p 247
[3] Yeliseyev Y, Savushkin B 2010 *EDM of aerospace products* (Moscow: MSTU him NE. Bauman) p 437
[4] Potapov V *High-speed milling and erosion processing: allies or rivals* (Electronic resource) URL: http://www.instr-stan.com/text/text_38.html
[5] Grisharin A, Ablyaz T and Ogleznev N 2017 *Improving the efficiency of electroerosive machining of parts of hydraulic cylinders and special-purpose products through the use of electrode tools with enhanced electroerosion properties* (Perm: Perm national research polytechnic university, University Press) p 151
[6] Avraamov Y and Shlyapin A 1999 *New composite materials based on immiscible components: production, structure, properties* (Moscow: MGIU) p 208
[7] Livshits A and Rosh A 1980 *EDM and electrochemical machining. Calculation, design, manufacture and use of electrode* (Moscow: Institute of Information on Mechanical Engineering) p 223
[8] Bains P, Payal H and Sidhu S 2017 *Analysis of Coefficient of Thermal Expansion and Thermal Conductivity of Bi-Modal SiC/A356 Composites Fabricated via Powder Metallurgy Route* doi:10.1115/ht2017-5122

IOP Conf. Series: Materials Science and Engineering **510** (2019) 012007 doi:10.1088/1757-899X/510/1/012007

# Optical method for formation of crystals in nanosuspension

**G Ivanova, V Khe and V Ivanov**

Physics and Theoretical Mechanics Department, Far Eastern State Transport University, 21 Seryshev Street, Khabarovsk, Russia

E-mail: galigd2015@yandex.ru

**Abstract.** It is proposed to use light pressure forces to form crystals from nanoparticles in a transparent nanosuspension. The paper discussed theoretical model of formation of crystal from nanoparticles on a bottom of cuvette by using the laser effect in liquid. The solution of one-dimensional task of the light induced mass transfer was received as depending on intensity of laser beam.

## 1. Introduction

Chemical methods for the formation of nanostructures (including photonic crystals) from the liquid phase are widely known. However, recent studies show a number of new opportunities. For example, Duhr and Braun show how the light-induced thermal diffusion of latex microparticles in a fluid leads to the formation of a two-dimensional crystal on a cooling surface in a convective fluid flow [1]. The concentration of microparticles increased in the zone of the light beam by several orders of magnitude. For nanoparticles, the realization of such method is difficult, since the coefficient of thermal diffusion is much less for them than for microparticles [2].

In this paper, it is proposed to use light pressure forces to form crystals from nanoparticles in a transparent nanosuspension.

## 2. Light-induced formation of crystals

Consider the liquid phase medium with the nanoparticles (dispersed phase) which is under the influence of the reference laser beam with a uniform intensity profile $I$ (figure 1).

Light pressure force acting on the nanoparticle from the high-power reference beam is equal to [3]:

$$F_p = \frac{128\pi^5 a^6 n_1}{3c_0\lambda^4}\left[\frac{m^2 - 1}{m^2 + 2}\right] I , \quad (1)$$

where $m = n_2 / n_1$, $n_1$, $n_2$ are the refractive substance indices of the dispersion medium and the dispersed phase respectively, $c_0$ is the velocity of light, $a$ is the particle radius, $\lambda$ is the light wavelength.

Balanced one-dimensional equation describing the dynamics of the concentration of the nanoparticles in a liquid phase medium with diffusion [3]:

$$\frac{\partial C}{\partial t} = D\nabla^2 C - V\nabla C . \quad (2)$$

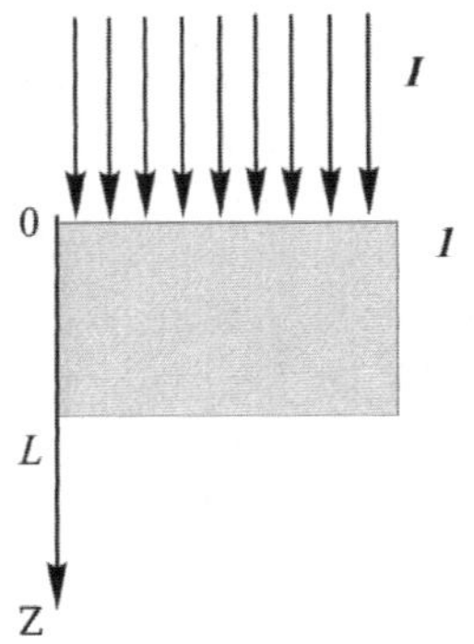


**Figure 1.** The light-induced sedimentation scheme: 1 – cell with liquid, $I$ – laser beam

here $C(z,t)$ is volume concentration of particulate matter, axis $z$ is aligned with the reference beam $I$, $D$ is diffusion coefficient; particle velocity $V = \left(64\pi^4 a^5 n_1 \left(m^2-1\right)\left(m^2+2\right)^{-1}\left(9c_0\lambda^4\eta\right)^{-1}\right)I$, where η is viscosity of the fluid.

The relevant boundary conditions:

$$-D\nabla C + \vec{V}C = 0\text{, when } z = 0 \text{ and } z = l, \tag{3}$$

where $l$ - the height of the cell along the propagation of the reference beam.

Initial conditions:

$$C = C_0\text{, when } t = 0, \tag{4}$$

where $C_0$ - the initial concentration of nanoparticles.

The exact solution of the equations (2) - (4):

$$C(z,t) = C_0\left\{1 - ue^{0.5uz}\sum_{m=1}^{\infty}\frac{2(\pi m)^2\left[1-(-1)^m e^{-0.5u}\right]}{\left[(\pi m)^2+u^2/4\right]^2}y_m\left[1-e^{-((\pi m)^2+u^2/4)\frac{D}{L^2}t}\right]\right\}, \tag{5}$$

where $y_m = (u/2\pi m)\sin(\pi mz) + \cos(\pi mz)$, $u = VL/D$.

In Figures 2 and 3 the solutions for $u$=3 and various intervals of times are shown. It is seen that the equilibrium actually occurs at $t = 1$.

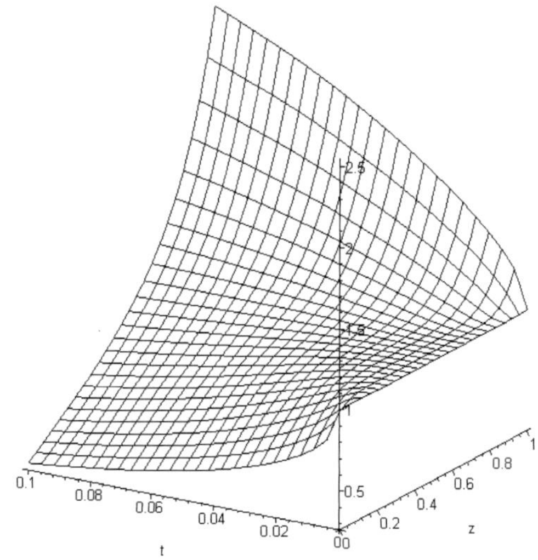


**Figure 2.** Function $C(z,t)/C_0$, if $u$=3 and $t \in [0;0.1]$

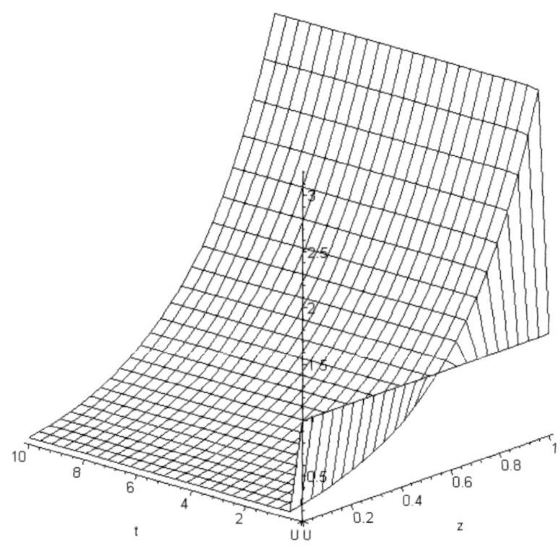


**Figure 3.** Function $C(z,t)/C_0$, if $u$=3 and $t \in [0;10]$

For the steady-state the expression (5) takes the next form:

$$C(z,I) = C_0 \gamma I D^{-1} l \frac{e^{\gamma I D^{-1} z}}{e^{\gamma I D^{-1} l} - 1}, \qquad (6)$$

where $\gamma = 64\pi^4 a^5 n_1 \left(m^2 - 1\right)\left(m^2 + 2\right)^{-1} \left(9 c_0 \lambda^4 \eta\right)^{-1}$.

The expression (6) we can write as:

$$C(z, l_{ph}) = C_0 l / l_{ph} \frac{e^{z/l_{ph}}}{e^{l/l_{ph}} - 1}, \qquad (7)$$

where we introduce parameter $l_{ph} = D / \gamma I$, that shows the depth at which the particle concentration changes by a factor of e at a given intensity.

The figure 4 illustrates the dependence of the parameter $l_{ph}$ on the radiation intensity for the following values: $a = 10^{-7}$ m, $n_1 = 1.1$, $n_2 = 1.33$, $\lambda = 632$ nm, $\eta = 1.004$ mPa·s.

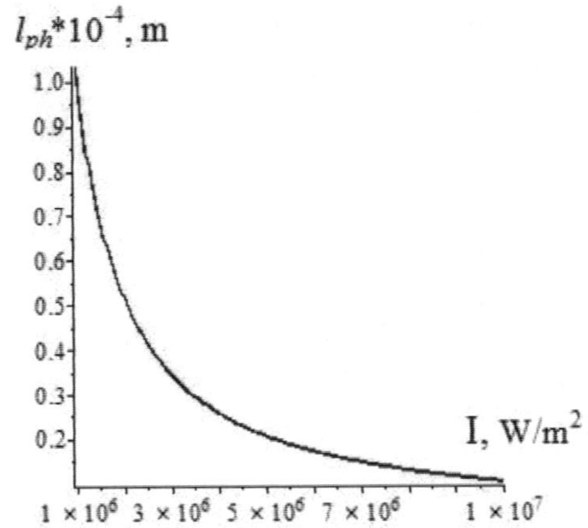


**Figure 4.** Dependence of the parameter $l_{ph}$ on the radiation intensity.

Also the expression (6) can be written as:

$$C_{r.u.}(z', I_{r.u.}) = I_{r.u.} \frac{\exp(I_{r.u.} z')}{\exp(I_{r.u.}) - 1}, \qquad (8)$$

where $C_{r.u.}(z', I_{r.u.}) = C(z', I) / C_0$ – relative concentration, $I_{r.u.} = I / I_{sat}$ – relative intensity of radiation, $I_{sat} = D / \gamma l$ – saturation intensity, $z' = z / l$.

## 3. Conclusions

Expression (8) gives a direct relationship between the change in the relative concentration and the relative intensity of radiation at the bottom of the cell. As estimates show, an essential increase in the relative concentration requires an intensity of about some MW/m$^2$, which is achievable for transparent nanomaterials using continuous laser sources. These light intensities form the conditions for the growth of the crystal structure.

Thus, an optical method for the formation of crystals from nanosuspension is proposed. A distinctive feature of the method is the use of light pressure forces, with the help of which it is possible to carry out effective sedimentation of nanoparticles in a transparent medium [4-6].

The proposed method for the formation of a colloidal crystal from dilute nanosuspension is also of interest for the formation of photonic crystals and chemical sensors [7-10]. The methods for efficient accumulation of nanoparticles from solution are of great interest for increasing the sensitivity of diffusion-limited surface biosensors [1]. The results can also be useful for the development of methods for optical diagnostics of nanomaterials [11-14].

**References**

[1] Duhr S and Braun D 2005 *Applied Physics Letters* **86** 131921
[2] Rabinovich G D 1981 *Separation of isotopes and other mixtures by thermal diffusion* (Moscow: Atomizdat) p 144 [in Russian]
[3] Ivanov V, Ivanova G, Krylov V and Khe V 2016 *Proceedings of SPIE* **10176** 1017607
[4] El-Ganainy R, Christodoulides D, Rotschild C and Segev M 2007 *Opt. Express* **15** 10207
[5] Buzzaccaro S, Tripodi A, Rusconi R, Vigolo D and Piazza R 2008 *Journal of Physics: Condensed Matter* **20** 494219
[6] Krylov V, Khe V and Ivanov V 2018 *Proceedings of SPIE* **10833** 108336N
[7] Ganguly S and Chakraborty S 2011 *Phys Lett A* **375(24)** 2394–2399
[8] Schuck P 2000 *Biophys. J.* **78(3)** 1606
[9] Ivanova G, Ovseychook O and Ivanov V 2018 *Proceedings of SPIE* **10833** 108334V
[10] Chintamani P, Shalini M, Agnel P, Meera V, Tejas I H and Radha S 2014 *International Journal of Chemical and Physical Sciences* **3**(5) 44
[11] Okishev K, Ivanov V, Kliment'ev S, Kuzin A and Livashvili A 2010 *Opt. Atm. and Ocean* **23** 106
[12] Tsukimura K, Suzuki M, Suzuki Y and Murakami T 2010 *Cryst Growth Des* **10(8)** 3596–3607
[13] Khe V, Ivanov V and Ivanova G 2018 *Proceedings of SPIE* **10833** 108336O
[14] Ivanov V and Ivanova G 2018 *Proceedings of SPIE* **10833** 108334U

IOP Conf. Series: Materials Science and Engineering **510** (2019) 012008 doi:10.1088/1757-899X/510/1/012008

# Light lens response in nanofluid

**G D Ivanova and V I Ivanov**

Department of Physics and Theoretical Mechanics, Far Eastern State Transport University, 47, Serysheva street, Khabarovsk, Russia

E-mail: tmeh@festu.khv.ru

**Abstract**. The light lens response in transparent nanosuspension with electrostrictive nonlinearity was analyzed. The theoretical analysis of the light-induced heat and mass transfer in the nanosuspension was carried out for the Gaussian beam radiation, when the concentration change gives rise to the thermal response due to the Dufour effect. The results are relevant to the study of the radiation self-action in the nanosuspension and optical diagnostics of such materials.

## 1. Introduction

Nonlinear optical methods are used successfully for the diagnostic of the nanomaterials [1-3]. The noncontact methods for determining the kinetic coefficients of such media are useful in the critical conditions (for example, near the phase transitions) and so they are very interesting instruments for investigators [3-5]. The light-induced thermal lens in a homogeneous fluid is formed as a result of thermal expansion of a medium. The optical nonlinear response of the nanosuspensions includes the parameters of the nanoparticles. The thermodiffusion effect (Soret effect) in the binary mixtures was experimentally investigated by the optical nonlinear methods in many works [6]. Another mechanism of optical nonlinearity of the medium is due to the forces operating on the particles of the dispersed phase in gradient light field [7]. This optical nonlinearity was studied experimentally in nanosuspensions and microemulsions [8-10].

The purpose of this work is the theoretical analysis of the contribution of the Dufour effect to the light induced lens response in the transparent nanosuspension with the electrostrictive nonlinearity.

## 2. Electrostrictive mechanisms of the cubic nonlinearity in nanosuspension

The coefficient of cubic nonlinearity for the electrostrictive mechanism is determined by the concentration nonlinearity:

$$n_2^{eff} = (dn/dC)(dC/dI) \tag{1}$$

Where $n$ is the refraction index of the medium, $I$ is the radiation intensity, $C$ is the concentration of nanoparticles, $n_2^{eff} = (dn/dI)$ is the coefficient of effective cubic nonlinearity.

In nanosuspension the particle radius is much smaller than the radiation wavelength $\lambda$, therefore the refractive index of the medium is proportional to the concentration of particles [4]:

$$n = n_1(1 + f\varepsilon), \tag{2}$$

where $\varepsilon = (n_2 - n_1)/n_1$ ; $n_1$ and $n_2$ are the refractive indices of the liquid and the dispersed phase, respectively; $f = v_0 C$ is the volume fraction of the dispersed phase, $a_0$ is the radius of the nanoparticle, $v_0 = (4/3)\pi a_0^3$ is the volume of the nanoparticle.

Balanced equation describing the dynamics of concentration of nanoparticles considering diffusion and electrostrictive flows can be written as [4]:

$$\frac{\partial C}{\partial t} = D\Delta C - div(\gamma C\nabla I), \tag{3}$$

where $D$ is the diffusion coefficient, $\gamma = 4\pi\beta D(\bar{c}nk_BT)^{-1}$, $\beta$ is the particle polarizability, $k_B$ is the Boltzman constant, $\bar{c}$ is the velocity of light in vacuum, $I(r)$ is the intensity of radiation.

The transparent nanosuspension is under the influence of laser radiation with Gaussian intensity profile [2]:

$$I = I_0 e^{-r^2/r_0^2}, \tag{4}$$

where $I_0$ is the radiation intensity on the beam axis, $r$ is the radial distance from the beam axis, $r_0$ is the beam radius.

This equation can be linearized in the case of low concentrations ($C \ll 1$):

$$(C(r,t) = C_0(1 + C'(r,t)),\ C'(r,t) \ll 1). \tag{5}$$

The exact solution of the equation (7) for this case can be written as [8]:

$$C(r,t) = C_0\left\{1 + \delta\frac{r_0^2}{4D}\left(\exp\left[-\frac{r^2}{r_0^2}\right] - \frac{r_0^2}{r_0^2 + 4Dt}\exp\left[-\frac{r^2}{r_0^2 + 4Dt}\right]\right)\right\}, \tag{6}$$

where $C_0$ is initial mass concentration of dispersed particles, $\delta = 4\tilde{I}_0\gamma r_0^{-2}$, $\tau_D = r_0^2/4D$ is the diffusion time.

## 3. Dufour effect in the Gaussian light field

Dufour effect describes a heat flow when there is a stream of one component of a binary system [5]. Induced radiation modification of particle concentration leads to heat flow. This kind of phenomena in linear non-equilibrium thermodynamics is called cross-effects and is described by the balanced equations [11]:

$$c_p\rho\frac{\partial T}{\partial t} = \lambda\nabla^2 T + \rho_1 T\mu_{11}^c S_T\left(\frac{\partial C}{\partial t}\right), \tag{7}$$

where $T$ is the fluid temperature, $c_p, \lambda$ are thermal constants, $\rho$ is the density of the material, $\mu_{11}^c = \partial\mu_1/\partial C$, $\mu_1$ is the chemical potential of dispersed phase, $S_T$ is Soret coefficient, $D$ is the diffusion coefficient.

The second term in equation (7) is related to that part of the heat flow, which occurs as a result of the particles flow in the system (Dufour effect).

Using equation (6) the heat problem is:

$$\frac{\partial T}{\partial t} = a\nabla_r^2 T + bC_0 \frac{\delta r_0^4}{(r_0^2 + 4Dt)^2}\left(1 - \frac{r^2}{r_0^2 + 4Dt}\right)\exp\left[-\frac{r^2}{r_0^2 + 4Dt}\right], \quad ((8)$$

$$T(r,0) = T_0\,, \quad \frac{\partial T}{\partial r}|_{r=0} = 0. \quad (9)$$

where $a = \lambda(c_p\rho)^{-1}$ is the thermal conductivity, $b = \rho_1 T \mu_{11}^c S_T C_0 (c_p \rho)^{-1}$.

The resulting expression describes the temperature dynamics of optically transparent nanofluid influenced by Gaussian beam:

$$T(r,t) = T_0 + C_0 \frac{b\delta r_0^4}{a}\left\{\frac{1}{r_0^2 + 4Dt}\exp\left[-\frac{r^2}{r_0^2 + 4Dt}\right] - \frac{1}{r_0^2 + 4at}\exp\left[-\frac{r^2}{r_0^2 + 4at}\right]\right\}. \quad (10)$$

where $\tau_{th} = r_0^2 / 4a$ is the thermal relaxation time.

It is easy to see that $T(r,t)$ as a function of time has a maximum. The appropriate time value $t_m$ in beam center ($r = 0$) is:

$$t_m = r_0^2 / \sqrt{aD}\,. \quad (11)$$

## 4. Light lens response

Let's take a two beam thermal lens scheme (see Figure 1). The reference (Gaussian) beam generates the thermal field in the optical cell with nanosuspension. The second (Gaussian) beam tests the light lens.

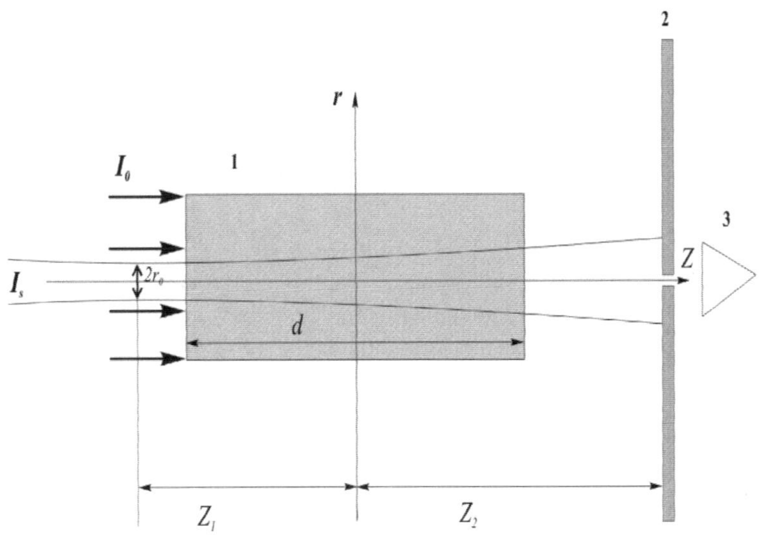


**Figure 1**. A light lens scheme (1 - the optical cell, 2- the screen with a pinhole, 3- the photodetector).

The light lens signal shows the change of the beam intensity $I(t)$ on the optical axis behind the screen:

$$\vartheta(t) = \frac{I(t) - I(0)}{I(0)} \tag{12}$$

where $I(0)$ is the beam intensity on the optical axis behind the screen at the starting point. The term for the lens transparency of the cell is used to calculate the light lens signal [5]:

$$\vartheta_{st} = -\frac{2(z_1/l_0)\Phi_{nl}(0)}{(1+z_1^2/l_0^2)(1+3z_1^2/l_0^2)} \tag{13}$$

Where $z_1, z_2$ are the distances from the cell centre to the Gaussian beam waist and to screen, respectively (fig. 1), $l_0 = \pi r_0^2/\lambda$ is the confocal parameter, $r_0$ is the radius of the Gaussian beam waist, $\Phi_{nl}(0)$ is the nonlinear phases in optical cell on the beam axis.

Nonlinear phase consists of two deposits: the first appears due to thermal expansion of the liquid phase and the second is associated with a change in the concentration of dispersed particles:

$$\Phi_{nl}(0) = \Phi_T(0) + \Phi_C(0) \tag{14}$$

$$\Phi_T(0) = k\int_0^d \left(\frac{\partial n}{\partial T}\right) \Delta T(z, r=0) dz \tag{15}$$

$$\Phi_C(0) = k\int_0^d \left(\frac{\partial n}{\partial C}\right) \Delta C(z, r=0) dz \tag{16}$$

Where $k = 2\pi/\lambda$ is the wave vector of the probing beam, $d$ is the thickness of a layer of the fluid.

The obtained results (6,10) allow us to calculate the summarized light induced lens response (for nonlinear phases):

$$\Phi_{nl} = n_1 d\delta\{f_0 D^{-1}[1-(1+4Dt/r_0^2)^{-1}] + (\partial n/\partial T)C_0 b a_0^{-1}(1+4Dt/r_0^2)^{-1}\} \tag{17}$$

where $(\partial n/\partial T)$ is the fluid parameter.

The last expression is received for the time $t >> \tau_{th}$. The second term is characterized by the negative sign and reduces the whole lens response. But this contribution disappears in the stationary state.

Thus, the expression was achieved for the light induced lens response in the nanosuspension.

## 5. Conclusions

The two-dimensional diffusion in the nanosuspension with electrostrictive nonlinearity in a Gaussian beam of radiation was analyzed taking into account the Dufour effect. The Dufour effect leads to the existence of the thermal contribution to the lens response in the transparent nanosuspension which disappears in the stationary state.

The expression was received for light induced lens response in nanosuspension. The results are relevant to nonlinear optics of nanosuspension [9-11], including the optical diagnostics of such nano-materials [12-15].

## References

[1] Freysz E, Afifi M, Ducasse A, Pouligny B and Lalanne J R 1984 *J. Opt. Soc. Amer. B* **1** 433

[2] Ivanov V I and Ivanova G D 2016 *Journal of Physics: Conference Series* **735** 012037
[3] Vicary L 2002 *Philos. Mag.* B. **82** 447-452
[4] Ivanov V I, Ivanova G D and Khe V K 2017 *Proc. SPIE* **10176** 1017607
[5] Malacarne L C, Astrath N G C, Medina A N, Herculano L S, Baesso M L, Pedreira P R B, Shen J, Wen Q, Michaelian K H and Fairbridge C 2011 *Optics Express* **19**(5) 4058
[6] Ivanov V I , Ivanova G D, Okishev K N and Khe V K 2016 *Proc. SPIE* **10035** 100354Y
[7] Lee W, El-Ganainy R, Christodoulides D, Dholakia K and Wright E 2009 *Optics Express* **17** 10277
[8] De Groot S R and Mazur P 1962 *Non-Equilibrium Thermodynamics* (Noth-Holland, Amsterdam) p 56
[9] Ivanov V, Ivanova G, Okishev K and Khe V 2017 *IOP Conf. Ser.: Mater. Sci. Eng.* **168** 012045
[10] Leppla C and Wiegand S 2003 *Philosoph. Mag.* **83** 1989
[11] El-Ganainy R, Christodoulides D, Rotschild C and Segev M 2007 *Opt. Express* **15** 10207
[12] Chintamani P, Shalini M, Agnel P, Meera V, Tejas I H and Radha S 2014 *International Journal of Chemical and Physical Sciences* **3**(5) 44
[13] Ovseychook O O, Ivanov V I and Ivanova G D 2018 *IOP Conf. Series: Journal of Physics: Conf. Series* **1038**, 012091
[14] Philip J and Nisha M 2010 *Journal of Physics: Conference Series* **214** 012035
[15] Kamanina V, Kuzhakov P V, Serov S V, Kukharchik A A, Petlitsyn A A, Barinov O V, Borkovskii M F, Kozhevnikov N M and Kajzar F 2013 *Proc. SPIE* **8622** 86221B

IOP Conf. Series: Materials Science and Engineering **510** (2019) 012009 doi:10.1088/1757-899X/510/1/012009

# Investigation of aluminum alloys by using subminiature eddy current probes

**S F Dmitriev[1], A V Ishkov[2], V N Malikov[1] and A O Katasonov[1]**

[1]Altai State University, Barnaul, Russia
[2]Altai State Agricultural University, Barnaul, Russia

E-mail: mirotnas@gmail.com

**Abstract**. On the basis of the eddy current probe(ECP) of a transformer type the measuring system that allows to search for defects in aluminum alloys is developed. The main technical data on the used eddy current probe and measuring system are given. The results of experiments obtained using different designs of the measuring system on two plates of duralumin D16T are presented. These dependences allow us to assess the quality of the alloys under study.

## 1. Introduction

Structures made from alloys based on aluminum are widespread in modern technology. The aluminum alloy (Al-Mg type) is widely used in structural applications, particularly in aerospace and automotive industry, due to the characteristics of good plasticity, good welding performance and corrosion resistance.

Conventional testing techniques such tension testing and micro-hardness testing, etc., are destructive measurement method. Some samples are taken from different parts of the object of control and are used for microstructure observation and property tests. However, microstructure and properties tested by the conventional approach just only present part of testing results of the component [1, 2]. In this regard, in modern industry, non-destructive methods are often used to test the object of control in the process of its control and operation [3,4]. On account of the fact that nondestructive testing doesn't damage the material performance, the change of physical parameters as ultrasonic sound velocity, attenuation coefficient and electrical conductivity are often used to deduce the variation of microstructure. Among many available non-destructive testing and evaluation (NDE) techniques ultrasonic test and eddy current test are widely studied to characterize the properties of the material. In recent decades, a number of studies of these methods have been carried out in order to evaluate properties of metals [4-6]. As a favorable method ultrasonic test has gained much attention in recent years, in investigating the relationship between the ultrasonic and the mechanical properties such as hardness, strength, impact strength, fracture toughness, etc. [7–10]. Eddy current testing is another important technique used to characterize materials, since eddy currents are influenced by microstructural alterations due to precipitates, cold work hardening, heat treatment, deformation, etc. Since Al-alloy is widely used for structural applications, therefore, extensive work has been conducted on characterizing properties by ultrasonic test and eddy current test. Therefore, the use of eddy current tests to study the properties of aluminum is relevant.

Eddy current (EC) testing is one of the most known and widely used NDE methods of conducting materials. The most common method to detect the presence of defects in conducting materials using EC is indirect, by registering the coil impedance variation due to the interaction between the applied AC magnetic field and the magnetic moment generated by eddy currents [11, 12]. Firstly, EC testing does not require an acoustic couplant in contrast to ultrasonic inspection. Secondly, EC is more economical and less hazardous than radiography. In addition, EC testing boasts many advantages over conventional eddy current testing, including more extended detection depth and easier generation and control. EC testing can be widely used to measure the thickness and stress [13] and to characterize crack, metal loss, and corrosion of metal materials and carbon fiber-reinforced plastic [14].

EC systems have been proposed in the last years due to the high sensitivity and wide frequency response of this type of transducers. An eddy current testing technique in nonmagnetic metals using a EC sensor and a pancake-type excitation coil was proposed some years ago [15]. The authors proved the advantages of the proposed method using two different hybrid probes, where the high sensitivity of the EC sensor is exploited by using excitation coils of just a few turns. They introduced sensors as a viable EC testing system, but they did not establish a correlation between the EC output voltage and the dimension of the defect, which could help in estimating its dimension. However, the correlation of the EC output voltage with a set of defects (having different width, depth, and length) was not considered, see for example [16–18]. An EC system to detect and characterize cracks and holes by artificial neural networks, using the same probe configuration proposed in [19]. The EC sensor was magnetically biased with the use of an AlNiCo permanent magnet, and the system was able to estimate only the flaw width and the bore diameter [19]. The authors in [16] proposed a novel design of rotating magnetic field eddy-current probe for non-destructive evaluation of non-ferromagnetic and ferromagnetic tubes in nuclear power plants and they analysed the influence of the geometrical parameters of the excitation coils and EC sensor arrays in the sensitivity of the probe but not the detectability of the proposed system and the quantification of defects intended to be detected with such probe.

Even though EC testing is a well-known and established technique in the industry, there are still a lot of efforts to find efficient models in order to estimate the crack's dimension and also to find the depth profile of narrow cracks [20-22]. These models are advantageous over their predecessors but the computational time still poses a constraint when quick estimation of crack dimensions is a necessity. For instance, the rapid computation method proposed in [20] takes approximately 0.2 s per inversion step to obtain EC crack signals.

An inverse problem algorithm was developed to determine the spatial eddy current distribution in the aluminum plates in order to get relevant information about complex geometry of cracks [23]. As expected, the results show good characterization of the geometry of surface cracks that are not parallel to the current flow. In [23], the same kind of probe was used to evaluate a linear crack with a set of different probe angles to overcome this difficulty. However, the method was slow due to the requirement of several scans. Furthermore, small rotation errors were detected when measuring the magnetic field map due to mechanical misalignment of the probe. For this purpose, a faster excitation method and an improvement to the probe were made to induce parallel currents in the specimen in different directions without physical rotation of the probe during the scan in order to improve the accuracy of the measurements.

## 2. Materials and methods

A superminiature ECP is designed to control the parameters of aluminum alloy products in small areas [24-27]. The obtained parameter is the electrical conductivity of the material and its distribution on the surface and thickness of the object of control. The developed measuring system works as follows:

The sinusoidal signal is generated by a sinusoidal voltage generator and comes to the exciting coil of the eddy current probe. In addition to the exciting coil, the probe also uses measuring and stabilizing coils connected differentially. Due to the differential inclusion of the coils, the minimum value of the initial unbalance of the probe is provided when it is removed from the controlled product. The measuring coil of the probe is connected to the amplifier with automatic gain control (AGC). AGC acts effectively in the range of permissible changes in the gap between the probe and the object of control. The signal from the output of the amplifier comes to one of the inputs of the amplitude detector, then passes the

IOP Conf. Series: Materials Science and Engineering **510** (2019) 012009 doi:10.1088/1757-899X/510/1/012009

filter, the analog-to-digital converter and is transmitted to a personal computer. The personal computer processes the output signal of the analog-to-digital converter and displays the information.

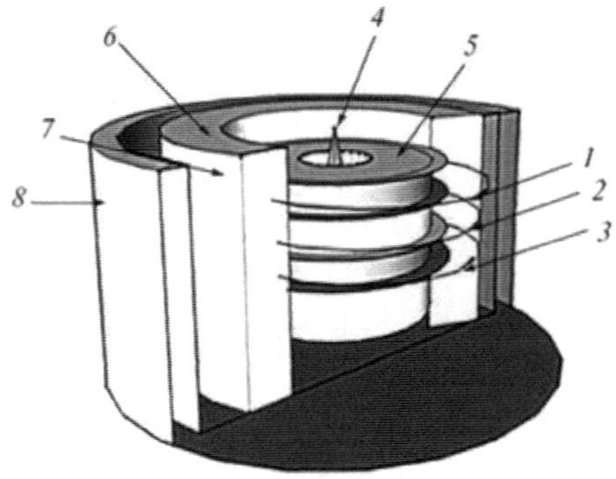


**Figure 1.** Eddy-current probe

The eddy current probe is a transformer with exciting, measuring and stabilizing coils placed on the core (Figure 1). The cores were made of ferrite 2000 NM3 with the value of initial magnetic permeability 500 and alloy 81NMA with the value of initial magnetic permeability 35000.
The characteristics of these ECPs allow for the localization of the magnetic field within the areas of 2500 $\mu m^2$ and, when using sufficiently low (100-700 Hz) frequencies on the exciting coil, allow to achieve a significant depth of its penetration into the object of control.

## 3. Experimental results and discussion

Experiments are conducted on D16T duralumin plates #1 and #2.

The first plate was 5.5 mm thick and contained three defects. The defects were slots imitating subsurface cracks. These slots were located at a depth of 1.3 and 4 mm. The width of slots was 1 mm. The second plate had a thickness of 5.5 mm and five slots with a width of 0.25 mm. The depth of slots was 1-5.3 mm.

In the study of the plate #1, the voltage on the exciting coil of the eddy current probe was 2 V.

In order to determine the sensitivity of the sensor to defects in the depth of the metal, scanning of the defect-free side of the sample was carried out.

The results of the study of plates are shown in Figures 2-5, Um – voltage on the measuring coil, l – distance from the edge of the plate to the sensor.

IOP Conf. Series: Materials Science and Engineering **510** (2019) 012009 doi:10.1088/1757-899X/510/1/012009

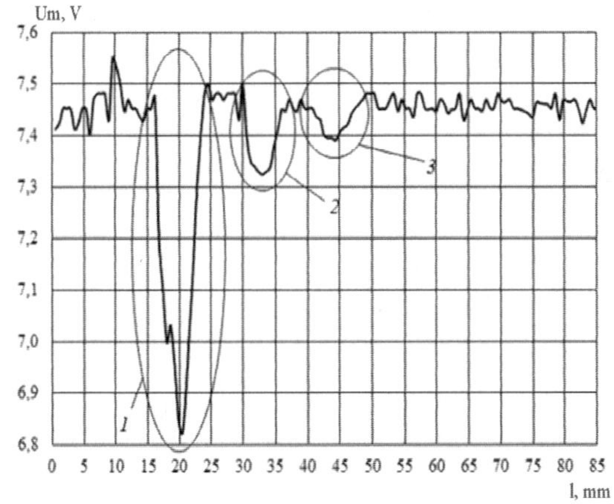


**Figure 2.** - Results of plates scanning #1 using a measuring system with one ECP (ferrite).

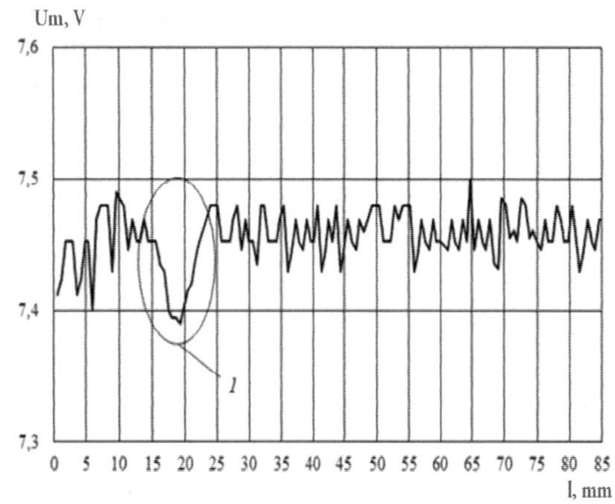


**Figure 3.** - Results of plates scanning #2 using a measuring system with one ECP (ferrite).

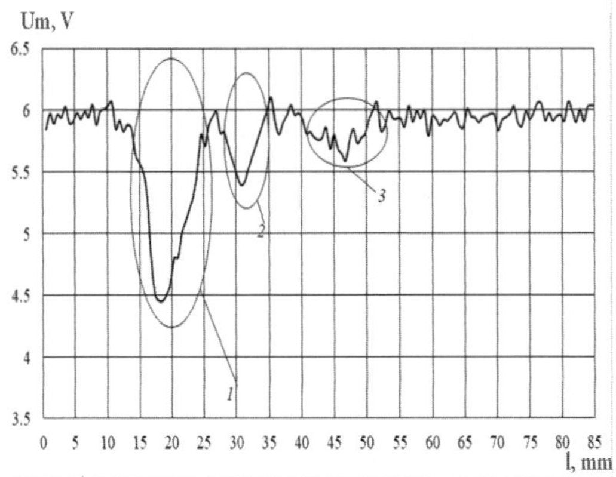


**Figure 4.** Results of plates scanning #2 using a system with two ECPs (ferrite).

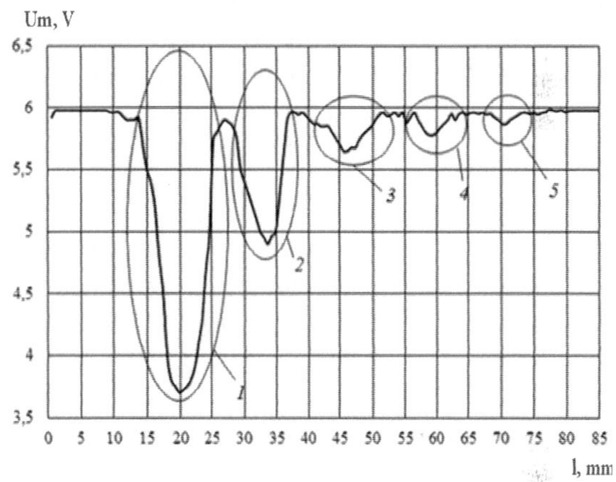


**Figure 5.** Results of plates scanning #2 using a system with two ECPs (81NMA).

The results of defectoscopy of the plate #1 with defects with a width of 1 mm at a frequency of ω = 500 Hz allowed to detect the slots to reduce the amplitude of the output signal Um. The decrease in the amplitude of the signal Um on the first defect was 0.75 V, on the second – 0.2 V, on the third – 0.1 V (Figure 2).

This study demonstrates the effectiveness of the developed eddy current sensor for finding defects of deep occurrence with a width of 1 mm and lying at a depth of 4 mm.

In order to determine the maximum localization of the magnetic field, the developed sensor was tested on the second plate with defects having a width of 0.25 mm, lying at a depth of 5.3 mm.

The measurements were carried out at a voltage of 2 V on the exciting coil. At the same time, it was managed to detect one of the defects occurring at a depth of 1 mm. The decrease in the amplitude of the output signal in this case was 0.1 V (Figure 3).

In order to improve the accuracy of measurements, an increase in the voltage on the exciting coil was made. In this case, there was an excess of the permissible value of the signal amplitude on the measuring coil and there was a significant distortion of the results.

In order to increase the localization of the magnetic field, the measuring system made on the basis of the subminiature eddy current probe was significantly upgraded as follows: the second eddy current probe was added to the developed hardware and software complex. At the same time, the first eddy current probe was static and located above the defect-free part of the sample. The second probe was used to scan the sample for defects. The signal of the second probe was subtracted from the signal received from the first one. Thus, the output signal was the voltage difference obtained from the measuring coil of both ECPs.

Defectoscopy of the sample #2 was performed at a frequency of 500 Hz. At the same time, it was possible to detect three defects out of six (Figure 4).

To further increase of localization of the magnetic field, it was necessary to introduce a probe made on the basis of a core of annealed alloy 81NMA with a magnetic permeability value of 35000 as a scanning probe. With this upgrade, it was possible to detect five defects out of six.

The decrease of the output signal amplitude Um on the first defect was 2.5 V (position 1), on the second defect – 1 V (position 2), on the third defect – 0.4 V (position 3), on the fourth – 0.2 V (position 4), on the fifth – 0.1 V (position 5). Changes in signal response when passing over the sixth defect are not fixed (Figure 5).

## 4. Conclusion
As a result of theoretical and experimental studies the eddy current measuring system that allows to search for defects in the alloy duralumin D16T was developed. The developed methods of signal processing allowed to increase the signal/noise relation of the eddy current probe by several times and significantly reduce the measurement error. Subminiature eddy-current transducers with specially shaped ferrite cores and modified filters can be used to localize a magnetic field on a small segment of a test object and to make the field produced by the exciter winding penetrate more deeply into the test object at a suitable frequency.

## References
[1] Jobba M, Mishrab K and Niewczasa M Int J Plast. 2015 **65** 43
[2] Mishina O, Godfrey A, Jensen D and Hansen K 2013 *Acta Mater.* **61** 6354.
[3] Zergoug M, Lebaili S, Boudjelial H 2004 *NDT E Int.* **37** 65
[4] Rajkumar V, Kumar A, Jayakumar T, Raj B and Ray K K 2014 *Metall Mater Trans.* **38A** 236.
[5] Badodo A, Lebaili S and Benchaala A 2016 *NDT E Int.* **36(1)** 1
[6] Vasudevan M and Palanichamy P 2002 *J Mater Eng Perform.* **11(2)** 169.
[7] Li A, Roberts R and Frank J M 2001 *Rev Prog Quant Nondestruct Eval* **80(11)** 2469.
[8] Lin L 2003 *Ultrasonic nondestructive characterization of microstructure of alloy steel* vol 1 (Dalian: Dalian University of Technology. Department of materials engineering). p 37.
[9] Nanekar P P and Shah B K 2003 *BARC Newsl.* **249** 25
[10] Kwun S I, Hong S T, Choo W Y 2000 *J Mater Sci Lett.* **19** 1453
[11] He Y, Simm A and Jackson P 2012 *IEEE Sensors Journal* **12** 2113
[12] Fan M and Zhou Z 2009 *NDT & E International* **42** 376
[13] Morozov M and Withers P J 2010 *Journal of Applied Physics* **108** 024909
[14] Wu J and Wang J 2014 *Journal of Sensors* **727** 269
[15] Dogaru T and Smith S T 2001 *IEEE Trans Magn.* **37(5)** 3831
[16] Jen-Tzong J, Guan-Shiun L, Wen-Chu L and Chia-Lun S 2016 *J Magn Magn Mater.* **304(1)** 470.
[17] Nair N V, Malpudi V R, Jimenez H R, Liu X, Deng Y and Zeng Z 2006 *IEEE Trans Magn.* **42(10)** 3312.
[18] Cacciola M, Megali M, Pllicano′ D, Morabito F C 2011 *Sensors Actuators A: Phys.* **167(1)** 25
[19] Postolache O, Geirinhas Ramos H and Lopes R A 2011 *Comput Stand Interfaces* **33(2)** 191
[20] Theodoulidis T, Poulakis N and Dragogias A. 2010 *NDT&E Int.* **43(1)** 13
[21] Theodoulidis T 2010 *NDT&E Int.* **43(7)** 591
[22] Theodoulidis T, Haitao W and Tian G 2012 *NDT&E Int.* **47(4)** 144
[23] Pasadas D J and Ramos H G 2015 Proc. IEEE Int. Instrum. Meas. Technol. Conf. (Pisa) vol 1 (Pisa: IEEE Xplore) p 940
[24] Malikov V N and Dmitiev S F 2016 *Russ. Eng. Res.* **32** 626

[25] Malikov V N and Dmitiev S F 2017 *Measurement Techniques* **60(4)** 372
[26] Dmitriev S F, Katasonov A O and Sagalakov A M 2017 *Materials Science Forum.* **927** 161
[27] Dmitriev S F, Katasonov A O, Sagalakov A M and Shevtsova L I 2015 *Journal of Physics: Conference Series* **643** 012058

IOP Conf. Series: Materials Science and Engineering **510** (2019) 012010 doi:10.1088/1757-899X/510/1/012010

# Development of a theoretical framework for investigation of compensation possibilities of deviations in the ship's piping system

**T M Do[1,2] and K N Sakhno[2]**

[1]Vietnam Maritime University, 484 Lach Tray, Le Chan, Hai Phong, Socialist Republic of Vietnam
[2]Astrakhan State Technical University, Astrakhan, Russian Federation

E-mail: dotatmanh@gmail.com

**Abstract.** The improving technology of ship piping systems at the design stage is an actual problem. The authors have analyzed technological features of manufacturing ship piping systems and have also considered the current state of design, manufacture and installation of ship pipeline issues. The review of research in the field of compensation of deviations of pipeline routes is presented. The problem of improving technology of ship piping systems at the design stage is considered providing the possibility of manufacturing pipes without measuring at the site. The authors have proposed the solution for the issue within studies of relationship of configuration and compensatory possibilities of pipeline routes. The authors are suggesting the idea of using straight pipes for moving the pipeline route to compensate for errors in the manufacture of pipes and the installation of tightly fixed connections of equipment, saturation products, etc., which ensures collecting the route without changing configuration of ready-made pipes. The proposed development allows to control and reduce at the design stage possible deviations that occur in the manufacture of pipes, to minimize the gaps assigned to the tracing of pipelines.

## 1. Introduction

A modern vessel is a sophisticated technological complex consisting of various types of equipment, mechanisms, structures, which operation provides with the piping systems (Figure 1). The labor intensity of pipeline production and ship system installation is up to 11-20% of the total manpower effort at the shipyard. The technology of pipeline production is characterized by a variety of operations and low level of mechanization. This production requires considerable and complex preparation [1-9].

**Figure 1.** Modeling of pipelines of ship systems with the help of computer program

In connection with the above, it is important to increase the efficiency of the pipeline production by introducing new technologies for the processing of pipes according to project documentation without measuring on the site [10, 11]. Availability of sufficient information for the production and installation of pipes in the project documentation allows to combine work on the vessel construction and shorten the time of shipbuilding orders execution [1,2,10,11].

## 2. Methods and Materials

To develop alternative methods of compensation, a hypothesis was put forward on the relationship between the configuration and compensation capabilities of the piping design trajectory. This hypothesis was substantiated in [10, 11]. Its main idea is to use straight and parallel sections with pipe connections to move the piping route. This compensates for the general deviations of rigidly fixed connections that limit this route, and also compensates for errors in the manufacture and installation of pipes, which will ensure the assembly of the route without changing the configuration of the finished pipes.

In the process of installation of the pipeline route, all deviations of the welded cups offset are compensated (Figure 2).

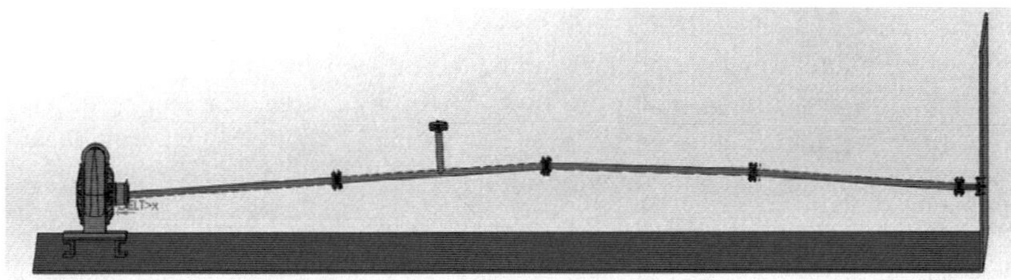

**Figure 2.** Pipeline corridor after deviation offset ΔX, ΔY, ΔZ – installation is effectively completed

In this study, we define the equations for determining the path compensation area in the direct problem (when designing the pipeline). In other words, we determine the possibility of compensating deviations of a given route. Consider the piping route, which consists of a sequence of points. Each of them can be either a point of connection of two tubes consisting of a sequence of points C ($x_C$, $y_C$, $z_C$), C ($x_{C1}$, $y_{C1}$, $z_{C1}$), ..., C ($x_{Cm}$, $y_{Cm}$, $z_{Cm}$), without changing the direction of the pipeline corridor, or the bending point F ($x_{F1}$, $y_{F1}$, $z_{F1}$), F ($x_{F2}$, $y_{F2}$, $z_{F2}$), ..., F ($x_{Fn}$, $y_{Fn}$, $z_{Fn}$) without changing the direction of the pipeline corridor (Figure 3). The last (terminate) point is A ($x_A$, $y_A$, $z_A$). The corridor must connect the start (C) and end point (A). The exact actual position of which relative to each other at the time of the corridor design is unknown.

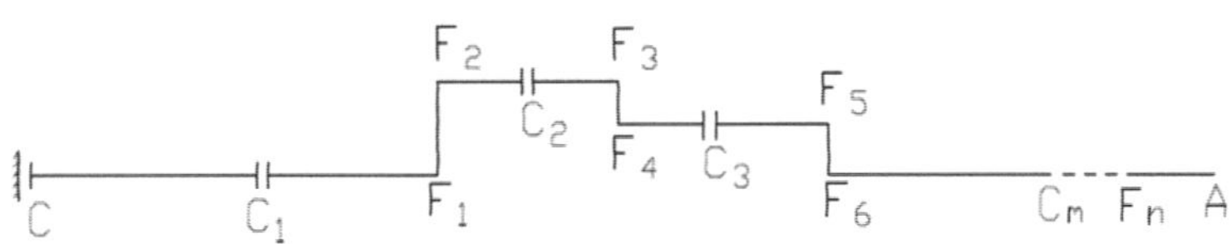


**Figure 3.** The route of pipe

The terminal point of the corridor describes the spherical segment while installing the first connection with the pipe squint, winding the corridor in the junction, found on these straight-through tubes (Figure 4). Further, while installing the second connection, non-perpendicular to the pipe axis, but parallel to each other, the terminate point of the pipeline corridor shifts under a certain trajectory. Likewise, while the axes of the following connections are shifted, the compensation area builds a three-dimensional body. This arc, surface and a three-dimensional body define the compensation area.

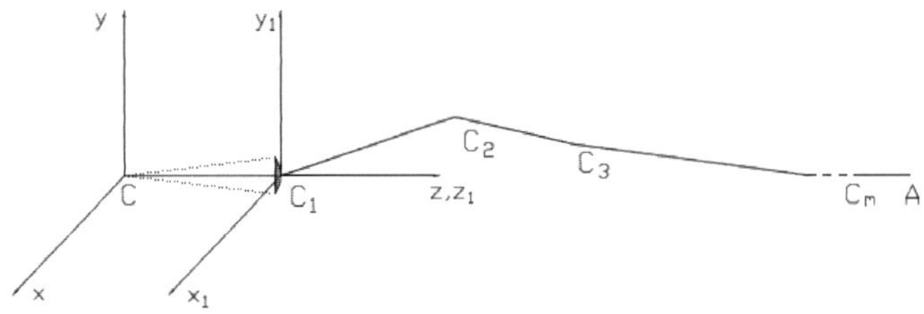


**Figure 4.** Transfer of system axes

The maximum deviation depends on the length of straight pipes or the distance between two connections. The research objective is to develop a mathematical description of the compensation area. For task, it is necessary to apply the calculation method based on the development of relevant theoretical concepts and mathematical formulas.

## 3. Results

We use the Cartesian coordinate system *Cxyz*. When installing connections at point C with a skew and rotation of the $CC_1$ , we obtain three coordinates ($r$, θ, φ), with $r$ – the shortest distance to the origin; θ, φ – zenith and azimuth angles, respectively. For task, it is more convenient to use spherical coordinate systems instead of the Cartesian ones.

Three coordinates (r, θ, φ) are defined as follows:

– $r \geq 0$ – distance from the origin of coordinates C to a given point;

– $0 \leq \theta \leq \alpha$ – the angle between the axis $Cz$ and the segment connecting the origin of the coordinates and the last point of the pipe $C_1$ (thus, $Cz$ – the pipe axis); α determined by OST 5. 95057-90 [12].

– $0 \leq \varphi \leq 2\pi$ – the angle between the X axis and the projection of the segment connecting the origin with the point $C_1$ on the Cxy plane (thus, Cxyz is the plane of the pipe).

If the spherical coordinates of the point $C_1$ are given, then the transition to the Cartesian Cxyz is carried out by the formulas:

$$\begin{cases} x_{C1} = r_1 \cdot \sin\theta_1 \cdot \cos\varphi_1; \\ y_{C1} = r_1 \cdot \sin\theta_1 \cdot \sin\varphi_1; \\ z_{C1} = r_1 \cdot \cos\theta_1, \end{cases} \quad (1)$$

где $r_1 = CC_1 = \sqrt{x_1^2 + y_1^2 + z_1^2} = \sqrt{(x_{C1} - x_C)^2 + (y_{C1} - y_C)^2 + (z_{C1} - z_C)^2}$ ; $\theta_1$ – skew angle [0, α]; $\varphi_1$ – angle of rotation [0, 2π].

Let's move back from Cartesian coordinates to spherical coordinates:

$$\begin{cases} r_1 = CC_1 = \sqrt{x_1^2 + y_1^2 + z_1^2} = \sqrt{(x_{C1} - x_C)^2 + (y_{C1} - y_C)^2 + (z_{C1} - z_C)^2} \\ \theta_1 = \arccos\left(\dfrac{z_1}{\sqrt{x_1^2 + y_1^2 + z_1^2}}\right) = \operatorname{arctg}\left(\dfrac{\sqrt{x_1^2 + y_1^2}}{z_1}\right) \\ \varphi_1 = \operatorname{arctg}\left(\dfrac{y_1}{x_1}\right) \end{cases}.$$

The system of equations of coordinates (1) determines the trajectory of moving point $C_1$. This trajectory is presented in Figure 5.

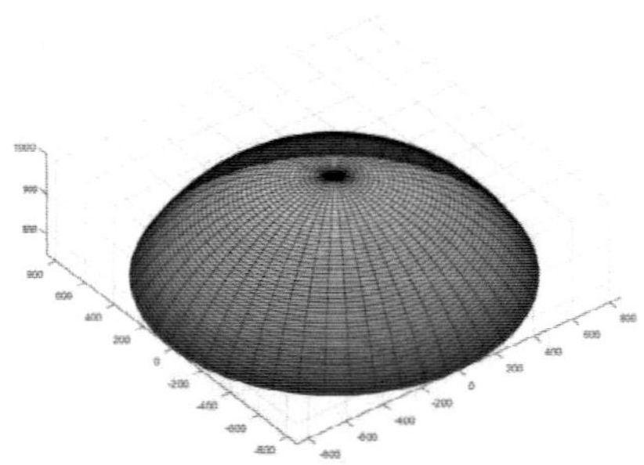

**Figure 5.** The trajectory of the point $C_1$ when turning the first pipe

With parallel transfer of the coordinate system Cxyz to point $C_1$, we will get a new coordinate system $C_1x_1y_1z_1$ (Figure 4). The formulas for the transition from Cxyz system to $C_1x_1y_1z_1$ system take the form:

$$\begin{cases} x = x_1 + x_{C1}; \\ y = y_1 + y_{C1}; \\ z = z_1 + z_{C1}. \end{cases} \quad (2)$$

In the new coordinate system $C_1x_1y_1z_1$, the coordinates of the connection point $C_2$ take the form:

$$\begin{cases} x_{C2} = r_2 \sin\theta_2 \cos\varphi_2; \\ y_{C2} = r_2 \sin\theta_2 \sin\varphi_2; \\ z_{C2} = r_2 \cos\theta_2, \end{cases} \quad (3)$$

with $r_2 = C_1C_2 = \sqrt{x_2^2 + y_2^2 + z_2^2} = \sqrt{(x_{C2} - x_{C1})^2 + (y_{C2} - y_{C1})^2 + (z_{C2} - z_{C1})^2}$;

Replace the systems of equations (1), (3), in the system of equations (2), the coordinates of the connection point $C_2$ take the form:

$$\begin{cases} x = x_{C2} + x_{C1}; \\ y = y_{C2} + y_{C1}; \\ z = z_{C2} + z_{C1}. \end{cases}$$

We get:

$$\begin{cases} x = r_1 \cdot \sin\theta_1 \cdot \cos\varphi_1 + r_2 \cdot \sin\theta_2 \cdot \cos\varphi_2; \\ y = r_1 \cdot \sin\theta_1 \cdot \sin\varphi_1 + r_2 \cdot \sin\theta_2 \cdot \sin\varphi_2; \\ z = r_1 \cdot \cos\theta_1 + r_2 \cdot \cos\theta_2. \end{cases}$$

Similarly, if the coordinate system of the following connection points is transferred in parallel (except for the last point A), then the coordinates of point A in the original coordinate system Cxyz take the form:

$$\begin{cases} x_A = \sum_{i=1}^{m} r_i \cdot \sin\theta_i \cdot \cos\varphi_i; \\ y_A = \sum_{i=1}^{m} r_i \cdot \sin\theta_i \cdot \sin\varphi_i; \\ z_A = \sum_{i=1}^{m} r_i \cdot \cos\theta_i, \end{cases} \quad (4)$$

with $r_i = C_{i-1}C_i = \sqrt{x_i^2 + y_i^2 + z_i^2} = \sqrt{(x_{Ci-1} - x_{Ci})^2 + (y_{Ci-1} - y_{Ci})^2 + (z_{Ci-1} - z_{Ci})^2}$; $m$ – quantity free connections in the pipe's route ($C$, $C_1$, $C_2$, …, $C_m$); $\theta_i$ – skew angle $[0, \alpha]$; $\varphi_i$ – angle of rotation $[0, 2\pi]$.

Thus, $x_A$, $y_A$, $z_A$ are the coordinates of the vector $\overrightarrow{CA}$. To determine the coordinates A after shifting the corridor on the junction in the original coordinate system Oxyz, use the formula:

$$\overrightarrow{OA} = \overrightarrow{OC} + \overrightarrow{CA}$$

Figure 6 shows two coordinate systems *OXYZ* and *Cxyz*:

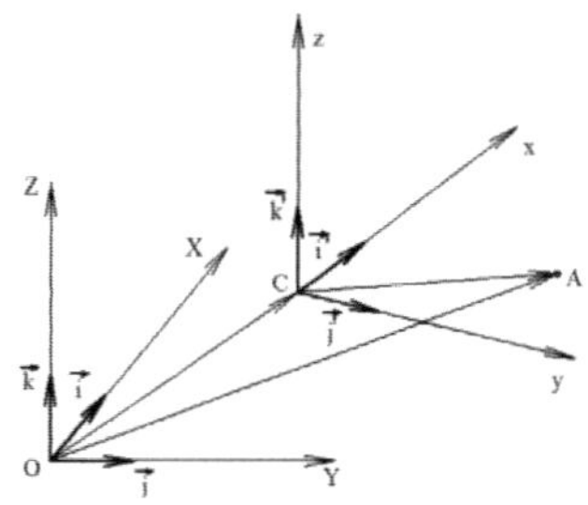


**Figure 6.** Communication between the Cxyz system and the original OXYZ system

For them:

– $\vec{i}$, $\vec{j}$, $\vec{k}$ – unit vectors directed along each axis X, Y, Z of system O;

– $\vec{i}'$, $\vec{j}'$, $\vec{k}'$ – unit vectors directed along each axis *x*, *y*, *z* of system *C*.

As shown above, $x_A$, $y_A$, $z_A$ – coordinates of point A in the system *Cxyz*; $X_A$, $Y_A$, $Z_A$ – coordinates of point A in the system *OXYZ*.

So in the original system *OXYZ*: $\overline{OA} = \vec{i}.X_A + \vec{j}.Y_A + \vec{k}.Z_A$; $\overline{OC} = \vec{i}.x_C + \vec{j}.y_C + \vec{k}.z_C$

In the system *Cxyz*: $\overline{CA} = \vec{i}'.x_A + \vec{j}'.y_A + \vec{k}'.z_A$

By the rule of addition of vectors, we have:

$$\overrightarrow{OA} = \overrightarrow{OC} + \overrightarrow{CA},$$

It means:

$$\vec{i}.X_A + \vec{j}.Y_A + \vec{k}.Z_A = \vec{i}.x_C + \vec{j}.y_C + \vec{k}.z_C + \vec{i}'.x_A + \vec{j}'.y_A + \vec{k}'.z_A \qquad (5)$$

Multiplying (4) by $\vec{i}$, we get:

$$(\vec{i}.\vec{i}).X_A + (\vec{i}.\vec{j}).Y_A + (\vec{i}.\vec{k}).Z_A = (\vec{i}.\vec{i}).x_C + (\vec{i}.\vec{j}).y_C + (\vec{i}.\vec{k}).z_C +$$
$$+ (\vec{i}.\vec{i}').x_A + (\vec{i}.\vec{j}').y_A + (\vec{i}.\vec{k}').z_A$$

According to the scalar product of vectors: $(\vec{i}.\vec{i}) = 1.1.\cos 0^o = 1$, $(\vec{i}.\vec{j}) = 1.1.\cos 90^o = 0$, $(\vec{i}.\vec{k}) = 1.1.\cos 90^o = 0$.

Formula (4) can be written as: $X_A = x_C + (\vec{i}.\vec{i}').x_A + (\vec{i}.\vec{j}').y_A + (\vec{i}.\vec{k}').z_A$

Similarly, multiplying (4) by $\vec{j}$ and $\vec{k}$ we get:

$$Y_A = y_C + (\vec{j}.\vec{i}').x_A + (\vec{j}.\vec{j}').y_A + (\vec{j}.\vec{k}').z_A$$
$$Z_A = z_C + (\vec{k}.\vec{i}').x_A + (\vec{k}.\vec{j}').y_A + (\vec{k}.\vec{k}').z_A$$

Determine the angles between $\vec{i}$, $\vec{j}$, $\vec{k}$ и $\vec{i}'$, $\vec{j}'$, $\vec{k}'$. Assign the axis of the first pipe $CC_1$ – *Cy*.

So: $\vec{j}' = \dfrac{\overrightarrow{CC_1}}{CC_1}$

This implies:

$$\cos(\vec{j}.\vec{j}') = \frac{y_{C_1} - y_C}{CC_1}. \qquad (6)$$

Because the vectors $\vec{i}$, $\vec{j}$, $\vec{k}$ and $\vec{i}'$, $\vec{j}'$, $\vec{k}'$ are mutually parallel or perpendicular, therefore, knowing the angle $(\vec{j}, \vec{j}')$, we can determine the remaining angles.

We obtain the coordinates of point A in the original OXYZ system:

$$\begin{cases} X_A = x_C + x_A; \\ Y_A = y_C + y_A; \\ Z_A = z_C + z_A; \end{cases}$$

or:

$$\begin{cases} x_A = x_C + \sum_{i=1}^{m} r_i.\sin\theta_i.\cos\varphi_i \\ y_A = y_C + \sum_{i=1}^{m} r_i.\cos\theta_i \\ z_A = z_C + \sum_{i=1}^{m} r_i.\sin\theta_i.\sin\varphi_i \end{cases} \qquad (7)$$

Equation (7) determines the trajectory of the displacement of the last point of trace A. This trajectory is shown in Figure 3.

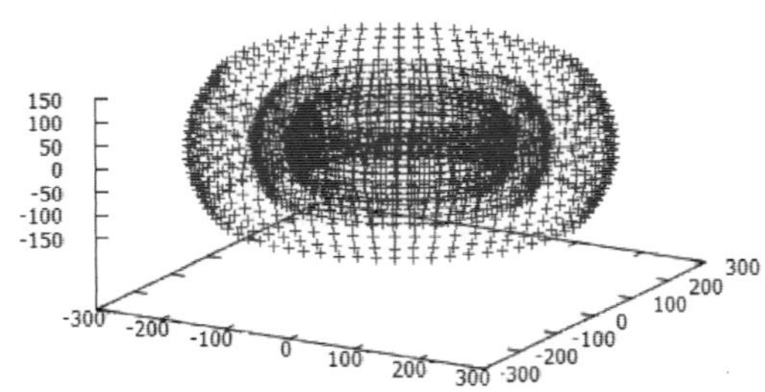


**Figure 7.** The trajectory of the displacement of point A when the route is rotated

## 4. Conclusions

To sum up the conducted studies, the following results were obtained: a mathematical description of the compensatory capabilities of pipeline routes was made using straight pipes; 3D models of the compensation area are constructed; the theoretical basis for the creation of an automated program is developed, which will allow determining the areas of compensatory opportunities for pipeline routes; it is possible to replace the measured pipes with the ready ones, which will help to reduce the time for building ships.

## References

[1] Safa M, Shahi A, Nahangi M, Haas C and Noori H 2015 *Structures* **3** 71-80
[2] Haiteng S, Wentie N, Yaxiao N, Chongkai Z and Weigao G 2016 *Computer Aided Drafting, Design and Manufacturing* **26(2)** 34–42
[3] Asmara A and Nienhuis U 2006 *Proceedings of the 5th International Conference on Computer and IT Applications in the Maritime Industries, Sieca Repro (TUD 06)* 269–280
[4] Park J H and Storch R L 2002 *Expert Systems with Application* **23(3)** 299–309
[5] Jiang W Y, Lin Y, Chen M and Yu Y Y 2015 *Ocean Engineering* **102** 63–70
[6] Lee D M, Kim S Y, Moon B Y and Kang G J 2013 *Journal of the Korean Society of Marine Engineering* **37(7)** 784–791
[7] Fan X N, Lin Y and Ji Z H 2007 *Shipbuilding of China* **48(1)** 82–90
[8] Ersoy M 2016 *Asymptotic Analysis* **98(3)** 237–255

[9] Zhang j Q and Liang Z S 2017 *International Journal of Applied Electromagnetics and Mechanics* **55(4)** 507–522
[10] Sakhno K N, Do T M and Vo Ch K 2016 *Mezhdunarodnyi nauchnyi zhurnal: Molodoi uchenyi. - Kazan* **6** 181-184
[11] Sakhno K N 2012 *PhD diss. Astrakhan': AGTU* 350
[12] OST 5.95057-90 1991 *L.: NPO «Ritm»* 207

IOP Conf. Series: Materials Science and Engineering **510** (2019) 012011 doi:10.1088/1757-899X/510/1/012011

# BEM modeling of a 3D homogeneous anisotropic elastic half space under dynamic load

**I P Markov and L A Igumnov**

Research Institute for Mechanics, National Research Lobachevsky State University of Nizhny Novgorod, 23, bldg. 6, Prospekt Gagarina (Gagarin Avenue), Nizhny Novgorod, 603950, Russia

E-mail: teanku@gmail.com

**Abstract.** In this paper, we consider a problem of a homogeneous anisotropic and linearly elastic half space subjected to dynamic loading. Zero body forces and vanishing initial conditions are assumed. The problem is solved using the Boundary Element Method in the Laplace transformed domain. Integral expressions of the three-dimensional dynamic fundamental solutions for displacements and tractions are utilized. We employ the displacement boundary integral equations which are regularized using the static part of the dynamic anisotropic traction fundamental solution. For the spatial discretization of the boundary integral equations mixed boundary elements are adopted. The geometry of the considered domain is approximated with quadrilateral quadratic eight-noded elements. On the boundary elements displacements and tractions are interpolated using linear and constant shape functions, respectively. Time-domain solutions are obtained using suitable scheme for numerical inverse Laplace transform. The boundary-element solutions for the illustrative problem of an anisotropic elastic half space subjected to a Heaviside-type load are provided.

## 1. Introduction

Modeling the elastic wave propagation in large unbounded domains is often required in a number of engineering problems such as dynamic soil-structure interaction, in the fields of geotechnical engineering, foundation engineering, seismology and so on. Particularly, Rayleigh surface wave propagation [1] is of significant importance. Most naturally occurring soils are rarely isotropic and in practical cases can be modelled by an anisotropic and linearly elastic medium. Taking into account anisotropy of the elastic properties brings considerable difficulties even for the relatively simple configurations of the problems involving unbounded domains and renders analytical examination of the dynamic wave propagation impossible. Thus numerical methods play the crucial role in studying the wave fields in infinite and semi-infinite anisotropic elastic regions.

Though the nature of the boundary element methods (BEM) makes them particularly advantageous for the problems involving infinite domains, formulations for anisotropic materials is rather scarce (see e.g. [2-8]). This is due to the unavailability of the explicit closed form of dynamic fundamental solutions (Green's functions) for anisotropic elastic media. Wang and Achenbach [9, 10] obtained integral expressions for singular (static) and regular (dynamic) parts of anisotropic elastodynamic Green's function. Evaluation of the regular term of fundamental solution requires numerical integration of the two-dimensional integral.

In this paper, to investigate a problem of the wave propagation in an anisotropic elastic semi-infinite region we employ a conventional direct boundary element formulation based on the regularized displacement boundary integral equations (BIEs) in Laplace domain. Spatial discretization of the BIEs is based on the idea of mixed boundary elements. The geometry of the considered domain is approximated with quadrilateral quadratic eight-noded elements. On the boundary elements displacements and tractions are interpolated using linear and constant shape functions, respectively. After obtaining results in the Laplace domain the time-domain solutions are obtained using modified Durbin's method for numerical inverse Laplace transform. The presented BEM formulation is applied to a 3D dynamic problem of anisotropic elastic half space subjected to a Heaviside-type traction.

## 2. Problem statement and BEM formulation

We consider a homogeneous, anisotropic and linearly elastic solid occupying volume $\Omega \subset R^3$ with boundary $\Gamma = \partial\Omega$. In the absence of the body forces and with vanishing initial conditions the Laplace-transformed equations of motion can be expressed as

$$C_{ijkl}\bar{u}_{k,lj}(\mathbf{x},s) = \rho s^2 \bar{u}_i(\mathbf{x},s), \quad \mathbf{x} \in \Omega, \tag{1}$$

where $\bar{u}_i$ is the displacement vector, $C_{ijkl}$ is the elastic stiffness tensor, $\rho$ is the mass density and $s$ is the Laplace transform parameter.

The boundary conditions are given as follows

$$\bar{u}_i(\mathbf{x},s) = \bar{u}_i^*(\mathbf{x},s), \quad \mathbf{x} \in \Gamma_u, \tag{2}$$

$$\bar{t}_i(\mathbf{x},s) = \bar{t}_i^*(\mathbf{x},s), \quad \mathbf{x} \in \Gamma_t, \tag{3}$$

where $\bar{t}_i$ denote the traction vector; $\bar{u}_i^*$ and $\bar{t}_i^*$ are the given displacements and tractions, correspondingly.

The regularized displacement boundary integral equations can be expressed as

$$\alpha\bar{u}_j(\mathbf{x},s) + \int_\Gamma \left[\bar{u}_k(\mathbf{y},s)\bar{h}_{jk}(\mathbf{r},s) - \bar{u}_k(\mathbf{x},s)h_{jk}^S(\mathbf{r}) - \bar{t}_k(\mathbf{y},s)\bar{g}_{jk}(\mathbf{r},s)\right]\mathrm{d}\Gamma(\mathbf{y}) = 0, \quad \mathbf{x} \in \Gamma, \tag{4}$$

where $\mathbf{r} = \mathbf{y} - \mathbf{x}$, $\bar{g}_{jk}(\mathbf{r},s)$ and $\bar{h}_{jk}(\mathbf{r},s)$ are the 3D displacement and traction fundamental solutions in Laplace domain, respectively; $h_{jk}^S(\mathbf{r})$ is the static term of $\bar{h}_{jk}(\mathbf{r},s)$.

Spatial discretization of the BIEs (4) is based on the idea of mixed boundary elements [11]. The quadrilateral quadratic elements are employed to approximate the geometry of the boundary $\Gamma$. The linear and constant variation of the displacements and tractions, respectively, is assumed over each element. After the nodal collocation procedure and some rearrangements we obtain a complex-valued set of linear algebraic equations for the fixed value of Laplace transform parameter $s$

$$\left[\bar{\mathbf{A}}(s)\right]\{\bar{\mathbf{p}}(s)\} = \{\bar{\mathbf{f}}(s)\}, \tag{5}$$

where $\{\bar{\mathbf{p}}(s)\}$ is the vector of unknown field variables, vector $\{\bar{\mathbf{f}}(s)\}$ contains prescribed boundary values and $\left[\bar{\mathbf{A}}(s)\right]$ is the system matrix.

## 3. Fundamental solutions

Laplace transformed dynamic fundamental solutions for anisotropic media can be represented as a sum of singular (static, superscript "$S$") and regular (dynamic, superscript "$R$") terms as follows

$$\bar{g}_{jk}(\mathbf{r},s) = g_{jk}^S(\mathbf{r}) + \bar{g}_{jk}^R(\mathbf{r},s), \tag{6}$$

$$\bar{h}_{jk}(\mathbf{r},s) = C_{ijkl}\bar{g}_{kp,l}(\mathbf{r},s)n_i(\mathbf{y}) = h^S_{jk}(\mathbf{r}) + \bar{h}^R_{jk}(\mathbf{r},s), \tag{7}$$

where $n_i(\mathbf{y})$ is the unit normal vector to the boundary at the point $\mathbf{y}$.

Singular and regular terms of the fundamental solution are expressed in the form of a one-dimensional integral over a unit circle and in the form of a two-dimensional integral over the surface of a unit hemi-sphere as follows [9, 10]

$$g^S_{ij}(\mathbf{r}) = \frac{1}{8\pi^2|\mathbf{r}|}\int\limits_{|\mathbf{d}|=1}\Gamma^{-1}_{ij}(\mathbf{d})dL(\mathbf{d}), \tag{8}$$

$$\bar{g}^R_{ij}(\mathbf{r},s) = -\frac{1}{8\pi^2}\int\limits_{\substack{|\mathbf{n}|=1\\ \mathbf{n}\cdot\mathbf{r}>0}}\sum_{m=1}^{3}\frac{k_m E_{im}E_{jm}}{\rho c_m^2}e^{-k_m|\mathbf{n}\cdot\mathbf{r}|}dS(\mathbf{n}), \tag{9}$$

with

$$c_m = \sqrt{\lambda_m/\rho},\ k_m = s/c_m,\ \Gamma_{ij}(\mathbf{d}) = C_{kijl}d_k d_l,\ \Gamma_{ij}(\mathbf{n}) = C_{kijl}n_k n_l, \tag{10}$$

$$dL(\mathbf{d}(\varphi)) \in D^S = \{0 \le \varphi \le 2\pi\},\ dS(\mathbf{n}(b,\varphi)) \in D^D = \{0 \le b \le 1; 0 \le \varphi \le 2\pi\}, \tag{11}$$

$$\mathbf{n}(b,\varphi) = \sqrt{1-b^2}\mathbf{d} + b\mathbf{e},\ \mathbf{e} = \mathbf{r}/|r|,\ \mathbf{e} = [e_1, e_2, e_3], \tag{12}$$

$$\mathbf{d}(\varphi) = \left[e_2\cos\varphi + e_1e_3\sin\varphi, -e_1\cos\varphi + e_2e_3\sin\varphi, -\left(1-e_3^2\right)\sin\varphi\right]\Big/\sqrt{1-e_3^2}. \tag{13}$$

where $\lambda_m$ are eigenvalues of $\Gamma_{jk}(\mathbf{n})$ and $E_{jm}$ are the corresponding eigenvectors.

## 4. Laplace transform inversion

The inverse Laplace transform is defined as the follows

$$f(t) = \frac{1}{2\pi i}\int\limits_{\alpha-i\infty}^{\alpha+i\infty}\bar{f}(s)e^{st}ds, \tag{14}$$

where $\alpha > 0$ is a real number greater than the real parts of all singularities of $\bar{f}(s)$.

It is impossible to directly use equation (14) in situations when values of $\bar{f}(s)$ can be computed only at the discrete set of values of the Laplace transform parameter $s$. Following the work of Durbin [12], with the known Laplace transform $\bar{f}(s)$ we have for the $f(t)$

$$s = \alpha + i\omega,\ f(0) = \frac{1}{\pi}\int\limits_0^\infty \mathrm{Re}\left[\bar{f}(\alpha+i\omega)\right]d\omega, \tag{15}$$

$$f(t) = \frac{e^{\alpha t}}{\pi}\int\limits_0^\infty\left\{\mathrm{Re}\left[\bar{f}(\alpha+i\omega)\right]\cos\omega t - \mathrm{Im}\left[\bar{f}(\alpha+i\omega)\right]\sin\omega t\right\}d\omega,\ t>0, \tag{16}$$

For more reliable and stable long time inversion we employ modification of the Durbin's method proposed by Zhao [13].

With $R$ being some large real number and defining $0 = \omega_1 < \omega_2 < ... < \omega_n < \omega_{n+1} = R$ we obtain

$$f(0) = \frac{1}{\pi}\sum_{k=1}^{n}\int\limits_{\omega_k}^{\omega_{k+1}}\mathrm{Re}\left[\bar{f}(\alpha+i\omega)\right]d\omega, \tag{17}$$

$$f(t)=\frac{e^{\alpha t}}{\pi}\sum_{k=1}^{n}\int_{\omega_k}^{\omega_{k+1}}\left\{\mathrm{Re}\left[\bar{f}(\alpha+i\omega)\right]\cos(\omega t)-\mathrm{Im}\left[\bar{f}(\alpha+i\omega)\right]\sin(\omega t)\right\}d\omega\ ,\ t>0. \quad (18)$$

On each interval $[\omega_k,\omega_{k+1}]$, $k=\overline{1,n}$, the $\mathrm{Re}\left[\bar{f}(s)\right]$ and $\mathrm{Im}\left[\bar{f}(s)\right]$ are approximated as follows

$$\mathrm{Re}[\bar{f}(\alpha+i\omega)]\approx F_k+\frac{1}{2}(Z_{k+1}+Z_k)(\omega-\tau_k)+\frac{1}{2\Delta_k}(Z_{k+1}-Z_k)(\omega-\tau_k)^2, \quad (19)$$

$$\mathrm{Im}[\bar{f}(\alpha+i\omega)]\approx F_k+\frac{1}{2}(Y_{k+1}+Y_k)(\omega-\tau_k)+\frac{1}{2\Delta_k}(Y_{k+1}-Y_k)(\omega-\tau_k)^2, \quad (20)$$

$$\Delta_k=\omega_{k+1}-\omega_k,\ \tau_k=(\omega_{k+1}+\omega_k)/2,\ F_k=\mathrm{Re}[\bar{f}(\alpha+i\tau_k)],\ G_k=\mathrm{Im}[\bar{f}(\alpha+i\tau_k)]. \quad (21)$$

The following system of linear algebraic equations is solved to determine $Z_k$ $k=\overline{1,n+1}$

$$\begin{cases}3\Delta_1 Z_1+\Delta_1 Z_2=8(F_1-F_0),\\ \dots\\ \Delta_{k-1}Z_{k-1}+3(\Delta_k+\Delta_{k-1})Z_k+\Delta_k Z_{k+1}=8(F_k-F_{k-1}),\\ \dots\\ 3\Delta_n Z_{n+1}+\Delta_n Z_n=8(F_{n+1}-F_n).\end{cases} \quad (22)$$

Likewise for the $Y_k$ we have the following system

$$\begin{cases}3\Delta_1 Y_1+\Delta_1 Y_2=8(G_1-G_0),\\ \dots\\ \Delta_{k-1}Y_{k-1}+3(\Delta_k+\Delta_{k-1})Y_k+\Delta_k Y_{k+1}=8(G_k-G_{k-1}),\\ \dots\\ 3\Delta_n Y_{n+1}+\Delta_n Y_n=8(G_{n+1}-G_n),\end{cases} \quad (23)$$

with $F_0=\mathrm{Re}[\bar{f}(\alpha+i\omega_1)]$, $F_{n+1}=\mathrm{Re}[\bar{f}(\alpha+i\omega_{n+1})]$, $G_0=\mathrm{Im}[\bar{f}(\alpha+i\omega_1)]$, $G_{n+1}=\mathrm{Im}[\bar{f}(\alpha+i\omega_{n+1})]$.

Finally,

$$f(0)\approx\frac{1}{\pi}\sum_{k=1}^{n}\left[F_k+\frac{\Delta_k}{24}(Z_{k+1}-Z_k)\right]\Delta_k, \quad (24)$$

$$f(t)\approx\frac{e^{\alpha t}}{\pi}\left\{\frac{1}{t}g_1(t)+\frac{1}{t^2}g_2(t)-\sum_{k=1}^{n}\frac{1}{t^3\Delta_k}\left[(Z_{k+1}-Z_k)(\sin\omega_{k+1}t-\sin\omega_k t)\right.\right.$$
$$\left.\left.+(Y_{k+1}-Y_k)(\cos\omega_{k+1}t-\cos\omega_k t)\right]\right\},\quad t>0, \quad (25)$$

where

$$g_1(t)=\left[F_n+\frac{1}{8}(Z_n+3Z_{n+1})\Delta_n\right]\sin\omega_{n+1}t+\frac{1}{8}(Y_2+3Y_1)\Delta_1-G_1$$
$$+\left[G_n+\frac{1}{8}(Y_n+3Y_{n+1})\Delta_n\right]\cos\omega_{n+1}t, \quad (26)$$

$$g_2(t)=-Z_1+Z_{n+1}\cos\omega_{n+1}t-Y_{n+1}\sin\omega_{n+1}t. \quad (27)$$

## 5. Numerical example

An anisotropic homogeneous elastic half space with mass density $\rho = 2216\ \text{kg/m}^3$ is considered. The otherwise traction free half space is subjected to a Heaviside-type tractions $t_3 = t_3^* H(t)$, $t_3^* = -1000$ Pa on a square with an area of 1 m² as depicted at figure 1. The elastic stiffness tensor of the half space is given in Voigt notation as follows [14]

$$\mathbf{C} = \begin{bmatrix} 17.77 & 3.78 & 3.76 & 0.24 & -0.28 & 0.03 \\ & 19.45 & 4.13 & -0.41 & 0.07 & 1.13 \\ & & 21.79 & -0.12 & 0.01 & 0.38 \\ & & & 8.30 & 0.66 & 0.06 \\ & \text{sym.} & & & 7.62 & 0.52 \\ & & & & & 7.77 \end{bmatrix} \text{GPa}. \quad (28)$$

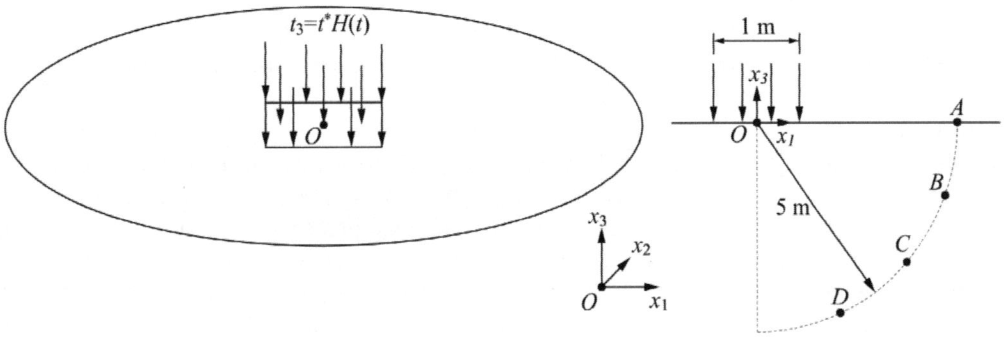


**Figure 1.** Configuration of the half space problem.

To use Zhao's modification of Durbin's method we need to define dimensionless frequencies $\omega_k$. To this end we employ the following relation

$$\omega_k = e^{(kx)^m} - 1, \quad (29)$$

with $m = 0.5$, $k = 1,2,...,800$ and $x = \left(\ln\left(\omega_{\max} + 1\right)\right)^{1/m} / k$, where $\omega_{\max} = 400$.

Displacements $u_1(t)$ and $u_3(t)$ at the surface point $A(5,0,0)$m and at the points inside half space $B(5\cos(\pi/8),0,-5\sin(\pi/8))$m, $C(5\cos(\pi/4),0,-5\sin(\pi/4))$m, $D(5\cos(3\pi/8),0,-5\sin(3\pi/8))$m, (see schematic representation at figure 1) are shown at the figures 2 and 3, respectively.

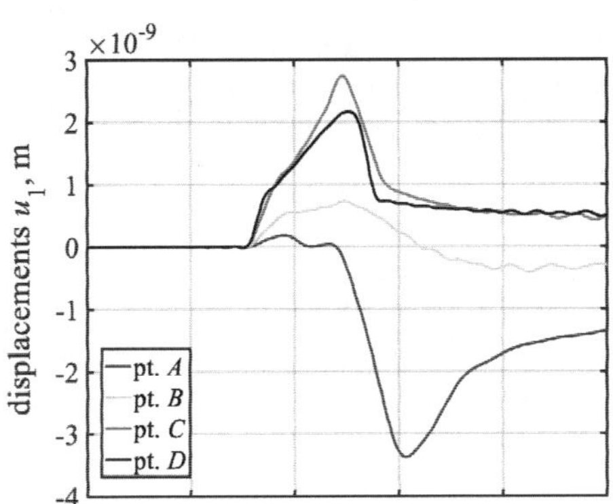


**Figure 2.** Displacements $u_1(t)$.

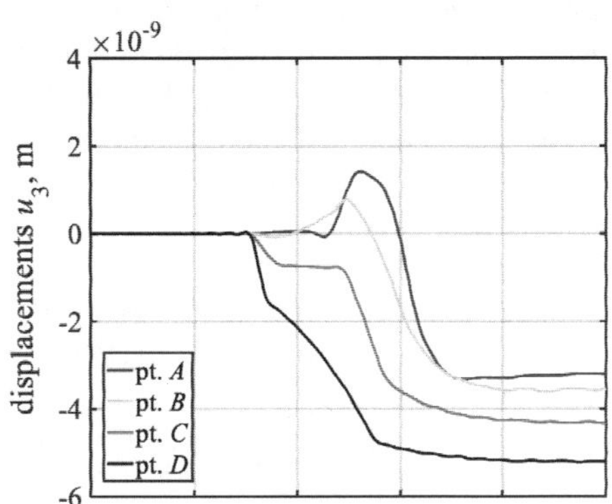


**Figure 3.** Displacements $u_3(t)$.

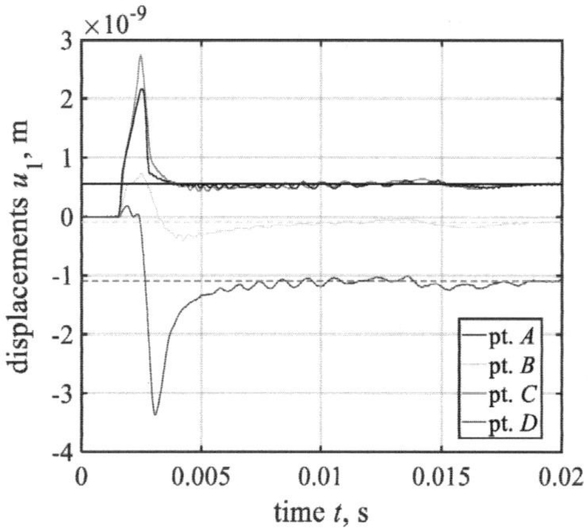


**Figure 4.** Displacements $u_1(t)$.

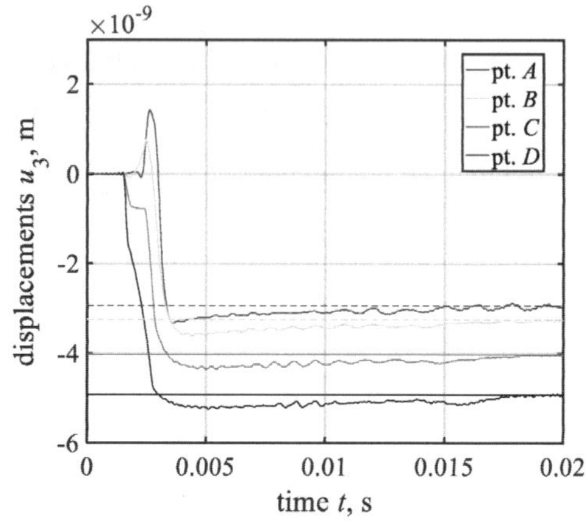


**Figure 5.** Displacements $u_3(t)$.

## 6. Conclusions

In this paper BEM formulation in the Laplace transformed domain was considered for problems of a homogeneous anisotropic and linearly elastic half space subjected to dynamic loading. Three-dimensional dynamic anisotropic fundamental solutions are represented as a sum of singular and regular terms which are expressed in the form of a one-dimensional integral over a unit circle and in the form of a two-dimensional integral over the surface of a unit hemi-sphere, respectively. Static term of the dynamic anisotropic traction fundamental solution is used to regularize the boundary integral equations. Spatial discretization rests on the idea of mixed boundary elements.

The described boundary element formulation is then applied to a dynamic problem of an anisotropic elastic half space subjected to a Heaviside-type tractions. Obtained results show that the presented Laplace domain boundary element approach coupled with the modified Durbin's method produce accurate and stable results for the dynamic problems involving the semi-infinite domains and anisotropic materials. On a larger time scale it can be observed that dynamic solutions at all points tend to their respective static counterparts as depicted at figure 4 for the displacements $u_1(t)$ and figure 5 for the displacements $u_3(t)$.

## Acknowledgments

The work is financially supported by the Russian Science Foundation under grant No. 18-79-00082.

## References

[1] Lamb H 1904 *Philos. Trans. R. Soc. Lond.* A **203** 1-42
[2] Ahmad S, Leyte F and Rajapakse R K N D 2001 *J. Eng. Mech.* **27** 149-156
[3] Chuhan Z, Yuntao R, Pekau O A and Feng J 2004 *J. Eng. Mech.* **130** 105-116
[4] Dravinski M and Wilson M S 2001 *Earthq. Eng. Struct. Dyn.* **30** 675-689
[5] Dravinski M 2003 *Earthq. Eng. Struct. Dyn.* **32** 653-670
[6] Niu Y and Dravinski M 2003 *Wave Motion* **38** 165-175
[7] Niu Y and Dravinski M 2003 *Int. J. Numer. Methods Eng.* **58** 979-998
[8] Furukawa A, Saitoh T and Hirose S 2014 *Eng. Anal. Bound. Elem.* **39** 64-74
[9] Wang C Y and Achenbach J D 1994 *Geophys. J. Int.* **118** 384-392
[10] Wang C Y and Achenbach J D 1995 *Proc. R. Soc.* A **449** 441-458
[11] Igumnov L A, Karelin I S and Petrov A N 2011 *Probl. Strength Plast.* **73** 97-103
[12] Durbin F 1974 *Comput. J.* **17** 371-376
[13] Zhao X 2004 *Int. J. Solids Struct.* **41** 3653-367
[14] Rasolofosaon P N J and Zinszner B E 2002 *Geophysics* **67** 230-240

IOP Conf. Series: Materials Science and Engineering **510** (2019) 012012 doi:10.1088/1757-899X/510/1/012012

# The effect of waste on the formation of cement clinker

**O A Miryuk**

Rudny Industrial Institute, Rudny 111500, Kazakhstan

E-mail: psm58@mail.ru

**Abstract.** The article is devoted to the development of resource-saving technology of Portland cement clinker using man-made raw materials. Wastes of skarn-magnetite ores enrichment were used in the experiments. Influence of the man-made component on the processes of clinker formation and composition of clinker phases were studied. Differential thermal and X – ray phase analysis methods were used to study the processes of formation and composition of clinker. It was revealed that the polymineral composition of the waste determines the intensity of low-temperature reactions, the variety of intermediate compounds. The formation of belite proceeds in three stages based on the structures of natural silicates (andradite, grossular, pyroxenes, amphiboles, chlorites, scapolite, epidote and feldspars). The bulk of the belite is formed at a temperature of 1050 – 1150 °C, this contributes to the intensification of the final stage of the formation of alite. Low-base clinker synthesized using ore dressing waste contains active modifications of alite and belite. Low-base cements (KN = 0.77 – 0.87) are characterized by increased strength. Man-made component provision into the raw mix reduces fuel consumption by 7.4 – 21.6 kg / ton of clinker. The research results prove the effectiveness of the use of technogenic materials for the production of cement clinker.

## 1. Introduction

The state of cement production determines the level of modern construction. Efficient development of the industry is promoted by saving raw and energy resources. Depletion of natural resources predetermines a fundamental change in the resource base due to the development of technogenic sources [1 – 15]. New raw materials differ from traditional materials in mineral composition [2, 7 – 11, 16]. Influence of mineral composition of technogenic raw materials on the cement clinker synthesis is not fully realized. Scientific ideas about consistent patterns of minerals influence of new raw materials on physicochemical transformations during formation of clinker phases are important in order to improve the efficiency of technogenic materials use.

The purpose of the work is to study the effect of technogenic raw materials on formation of cement clinkers and their properties.

## 2. Materials and methods of research

When enriching skarn-magnetite ores, waste of dry magnetic separation (DMS) is released, which is being crushed stone with a grain size of up to 25 mm. DMS waste has a stable chemical composition, wt.%: $SiO_2$ 40 – 45; $Al_2O_3$ 10 – 12; $Fe_2O_3$ 15 – 17; CaO 12 – 13; MgO 5 – 6; S 2 – 3; $R_2O$ 2 – 4; $TiO_2$ 0.50 – 0.53; $P_2O_5$ 0.25 – 0.30; MnO 0.35 – 0.40; $V_2O_5$ 0.04 – 0.06; Cl 0.09 – 0.12; Cu 0.04 – 0,05; Ni 0.007 – 0.008. The mineral base is formed by silicates, differing in genesis, composition, structure, physical properties, chemical activity and thermal stability, wt.%: pyroxenes (diopside) 20 – 25; epidote 10 – 13; feldspars 8 – 12; chlorites 7 – 10; scapolite 8 – 11; garnets (andradite, grossular) 7 – 12; amphiboles (actinolite) 7 – 14.

Waste also contains minerals, wt.%: calcite 4 – 7; pyrite 4 – 8; quartz 2 – 4; magnetite 3 – 4.

Pyrogenic transformations in a raw mixture were studied with the differential thermal analysis method, which was carried out on a derivatograph of the Q –1500 D of system F. Paulik, J. Paulik, L. Erdey. The phase composition of the raw mixture, burnt at temperatures of 500 – 1400°C, was determined using general purpose X – ray diffractometer, type DRON – 3M. The diffractometer is equipped with an BSV – 24 type X – ray tube with CuK α – radiation. The diffractograms were processed using difWin program. Strength properties were determined on samples with the size of 20x20x20 mm, made with dough of normal density.

## 3. Results and discussion

Influence of technogenic material on clinker formation processes was studied using the example of a raw material mixture of 70% limestone and 30% ore dressing waste.

### *3.1 The influence of technogenic material on clinker formation processes*

Burning of the studied raw material mixture is accompanied by numerous transformations (figure 1). Low-temperature heat release (exothermic effect at 500°C) is associated with the oxidation of pyrite and magnetite. Exothermic oxidative processes intensify decarbonization reactions and the decay of unstable minerals. Interaction of components begins at low temperatures due to the Hedvall effect during the period of structural transformations of minerals. Low-temperature melting of the technogenic component until appearance of a clinker melt ensures formation of a low-viscosity liquid phase and contributes to the activation of phase transformations.

The difference in crystalline structures of natural silicates and intermediate silica-containing phases predetermined the multi-stage bleaching with the predominant formation of $C_2S$ in the high-temperature region. The number of stages corresponds to the number of exothermic effects on the thermogram (figure 1).

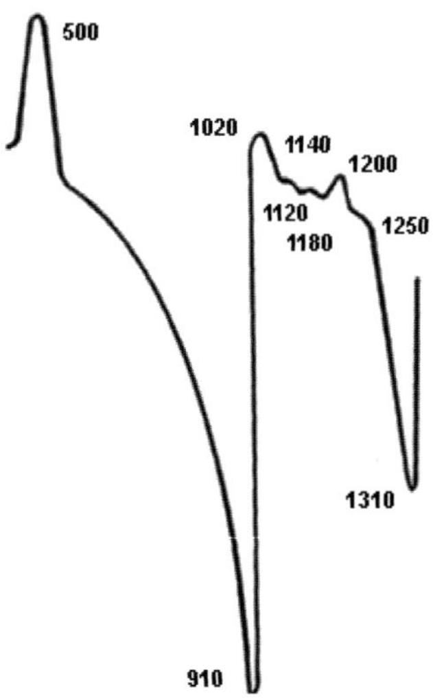


**Figure 1.** Thermogram of the raw material mixture «limestone – ore dressing waste».

In the solid-phase region (temperature 850 – 1050°C), dicalcium silicate is formed with the participation of the source minerals as a result of the introduction of excess amounts of calcium oxide into structure of garnets, scapolite [17].

At the next stage (temperature 1050 – 1150°C), it is formed by decomposition of intermediate phases and minerals of the technogenic component. The activity of belite formation in this temperature range is due to appearance of a sufficient amount of melt with a low viscosity.

The third stage (temperature 1150 – 1250°C) is characterized by formation of a belite based on aluminum silicates of calcium and sodium; magnesium silicates of calcium and aluminum, which were formed on the basis of natural silicates.

The main mass of the belite in the mixture under study is formed in the last two stages. The shift of the whitening reactions to the high-temperature region provides an increase in the heat stress during sintering the mixture. This creates prerequisites for the active formation of clinker phases and improvement of the crystal structure of clinker.

The presence on the thermogram of a double maximum at temperatures of 1250 and 1310°C is caused by a change in the rate of formation of alite. Alite formation, initially suppressed by influence of alkalis and sulphates, is accelerated when the melt is enriched with magnesium oxide and intensified due to high-temperature belite formation and dopants [16].

Increased temperature of belite formation, presence of dopants in mixtures with waste ensure the activity of low-base clinkers (KN = 0.73 – 0.83) [17 – 20]. Formation of belite at elevated temperatures prevents passivation of $C_2S$, increases doping of the phases of clinker. In clinker, from a technogenic raw material the phases inherit the features of the crystalline structures of natural minerals, which are characterized by close contact of various elements.

X – ray phase analysis of low-base clinkers determined modification of silicate clinker phases (figure 2).

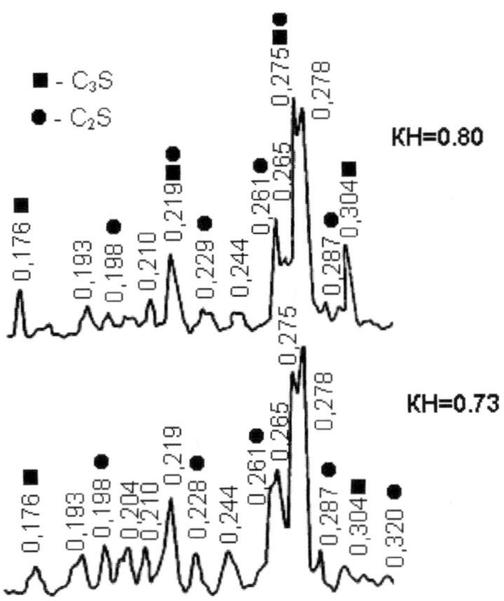


**Figure 2.** X – rays of clinkers on the basis of ore dressing waste.

Dicalcium silicate is represented by β-modification (d = 0.287; 0.278; 0.219 nm), which is stabilized by an increased concentration of $SO_3$ in the clinker. Presence on the radiograph d = 0.320; 0.278; 0.275 nm indicates the presence in the clinker of α' – $C_2S$, stabilized with magnesium oxide, manganese impurities. According to configuration of the maximum d = 0.176 nm, modification of the alite was determined: rhombohedral $C_3S$ (KN = 0.80) and monoclinic $C_3S$ (KN = 0.73).

*3.2 Influence of technogenic material on fuel consumption during clinker burning*

To determine the effect of technogenic material on properties of cements, clinkers were synthesized on the basis of raw mixtures with different content of DMS waste. In mixtures of the G series, waste functions as an iron-containing and aluminosilicate component. Mixes of the F series were obtained by replacing DMS in the control S series of the glandular component only (table 1). The resulting cements are characterized by high strength properties. Low-base cements (KN = 0.77 – 0.87) are characterized by increased hydraulic activity.

Efficiency of introduction of the technogenic component is determined from theoretical heat consumption for physicochemical transformations in the raw mix. The thermal effect of clinker formation was calculated as an algebraic sum of the heat of endothermic processes (decarbonization of limestone, dissociation of minerals, formation of a liquid phase) and exothermic reactions (pyrite oxidation, formation of clinker phases).

**Table 1.** Cement strength characteristics.

| Cement series | KN | Modules | | Strength at compression, MPa, at age, days | | |
|---|---|---|---|---|---|---|
| | | *n* | *p* | *2* | *7* | *28* |
| **G – 1** | 0.95 | | 0.9 | 95 | 103 | 125 |
| **G – 2** | 0.87 | | 0.9 | 83 | 90 | 110 |
| **G – 3** | 0.77 | | 0.9 | 50 | 70 | 105 |
| **F – 1** | 0.95 | | 1.2 | 101 | 108 | 118 |
| **F – 2** | 0.87 | 2.5 | 1.2 | 81 | 95 | 110 |
| **F – 3** | 0.77 | | 1.2 | 54 | 73 | 100 |
| **S – 1** | 0.95 | | 1.2 | 78 | 85 | 110 |
| **S – 2** | 0.87 | | 1.2 | 77 | 82 | 110 |
| **S – 3** | 0.77 | | 1.2 | 48 | 60 | 95 |
| Control cement M550 | | | | 90 | 108 | 120 |

Presence of calcium and magnesium silicates in waste contributes to a reduction by 4 – 11% of the thermal costs of decarbonization in mixtures with the same KN value. Saving heat on dissociation of carbonates is also achieved by KN changing from 0.95 to 0.77; for mixtures of the G, F, and S series, the savings are 165, 151 and 116 kJ/kg clinker, respectively. The theoretical heat consumption for the highly basic mixtures G – 1 and F – 1 is less than for the mixture of low basicity S – 3 from traditional raw materials (figure 3).

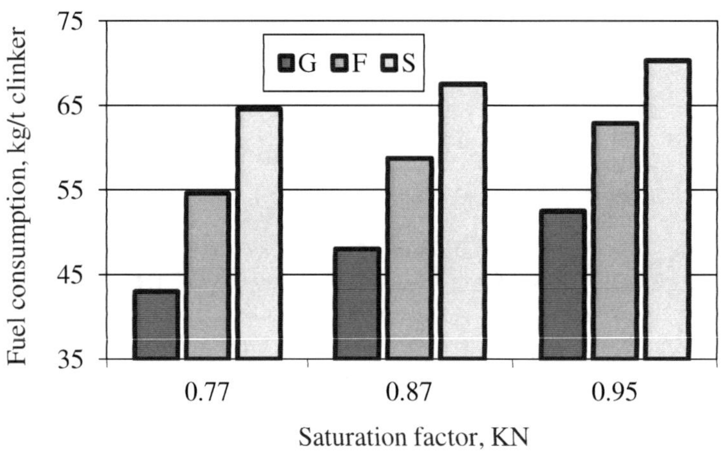


**Figure 3.** Specific fuel consumption for clinker formation when burning raw mixes with different content of waste DMS.

As clay is removed from the raw material mixture, thermal costs of dehydrating minerals are reduced, and the savings reach 47 – 150 kJ/kg clinker. A large part of heat consumption in the raw material mixtures with waste compensates heat released during the oxidation of pyrite. Energy benefit from presence of pyrite in the mixtures with DMS waste is comparable to or exceeds that from a decrease in KN for mixtures of compositions S – 1 and S – 3. The efficiency of mixtures with waste illustrates change in specific fuel consumption for clinker formation (figure 3). With introduction of the raw material mixture of technogenic component fuel economy is 7.4 – 21.6 kg/t clinker.

## 4. Conclusions

The use of ore dressing wastes to produce cement clinker minimizes natural materials content in the raw mix. With participation of the technogenic component, clinker formation is of a stepped nature, characterized by activity of low-temperature interactions, complex transformations of minerals, and a variety of intermediate phases.

Polymineral composition of ore dressing waste contributes to the formation of low-power clinker with high hydraulic activity. Reducing carbonate and fuel consumption will decrease harmful effects of production emissions.

Therefore, production of cement clinker using technogenic raw materials is technically feasible, cost-effective and environmentally preferable.

## References

[1] Cao H, Liu W, Xu J, Liu J, Huang J, Huang X and Li G 2018 *Water science and technology* **77** 960
[2] Oproiu C, Voicu G, Nicoara A I and Badanoiu A I 2018 *Revista de chimia* **69** 860
[3] Li H, Huang Y, Yang X J, Jiang Z and Yang Z H 2018 *Journal of material cycles and waste management* **20** 1701
[4] Gartner E and Sui T 2018 *Cement and Concrete Research* **114** 27
[5] Shettima A U, Hussin M W, Ahmad Y and Mirza J 2016 *Construction and Building Materials* **120** 72
[6] Buruberri L H, Seabra M P and Labrincha J A 2015 *Journal of Hazardous Materials* **286** 252
[7] Esteves T C, Rajamma R, Soares D, Silva A S, Ferreira V M and Labrincha J A 2012 *Construction and Building Materials* **26** 687
[8] Schoon J, De Buysser K, Van Driessche I and De Belie N 2013 *Construction and Building Materials* **48** 725
[9] Kairakbaev A K, Abdrakhimova E S and Abdrakhimov V Z 2016 *Glass and Ceramics* **73** 266
[10] Hojamberdiev M, Muhamedbaeva Z and Madhusoodana C D 2009 *Construction and Building Materials* **23** 2757
[11] Damtoft J S, Lukasik J, Herfort D, Sorrentino D and Gartner E M 2008 *Cement and Concrete Research* **38** 115
[12] Schoon J, Van der Heyden L, Eloy P, Gaigneux E M, De Buysser K and Van Driessche I 2012 *Construction and Building Materials* **36** 391
[13] Torkittkul P, Chaipanich A 2010 *Cement and Concrete Composites* **32** 440
[14] Pardo N, Moya J A and Mercier A 2011 *Energy* **36** 3244
[15] Schoon J, De Buysser K, Van Driessche I and De Belie N 2015 *Cement and Concrete Composites* **58** 70
[16] Li X, Huang H, Xu J, Ma S and Shen X 2012 *Construction and Building Materials* **37** 548
[17] Miryuk O A and Luginina I G 2001 *Russian Journal of Applied Chemistry* **74** 1789
[18] Chatterjee A K 1996 *Cement and Concrete Research* **26** 1213
[19] Kim Y M and Hong S H 2004 *Journal of the American Ceramic Society* **87** 900
[20] Kacimi L, Simon-Masseron A, Salem S, Ghomari A and Derriche Z 2009 *Cement and Concrete Research* **39** 559

# Special features of fused welding flux granular forming in carbon steel surfacing during plasma granulation

**S V Naumov, A O Artemov, M N Ignatov and K I Belousov**

Perm National Research Polytechnic University, 29, Komsomolsky prospekt, 614990, Perm, Russian Federation

E-mail: NaumovStanislav@yandex.ru

**Abstract.** This paper proposes a qualitatively new approach for production fused welding fluxes in carbon steel surfacing based on direct-action plasma arc influence on a metal substrate and charge materials made from the Ural region raw minerals. During plasma granulation there is an intense burnout of alkali metal oxides that is compensated by the use of liquid potassium and / or sodium glass as a binder of charge materials before granulation. Forming a slag crust on granules surface resulted in iron content exceeding more than 90 wt. % in the granule center. Otherwise without the formation of a slag on granules surface the iron was oxidized and its content reached 38.42 wt. % Studies have shown that the volume of granules that are completely coated and protected by slag crust on an oxide-silicate basis, reaches up to 95%. Such welding flux granules with a high content of metal components provide high performance (deposition rate) during repair works and surfacing of products using submerged arc welding.

## 1. Introduction

Welding materials are an essential part of modern welding production. Mineral-based welding materials are particularly in demand, namely, coated electrodes, flux-cored wires, welding fluxes of different purpose. Using high-quality welding materials with high welding and technological properties is required during welding and repairing (surfacing) of critical structures and general-purpose building structures. To this end, chemical composition of welding materials should be conducive to the stability of the arc and favorable process of physical and chemical reactions in welding (surfacing). Manufacture methods of welding materials are also being improved to reduce energy and material costs and increase welding properties on the whole [1-3].

Currently, plasma technologies for processing materials are increasingly used in various industries, primarily due to a highly concentrated source of heat, equipment simplicity and easy readjustment for various tasks and cost-effectiveness of the process while receiving final high-quality products [4].

Plasma technologies of processing and production of powders of various dispersity from rutile, iron oxide, clay and other mineral raw materials are proved to be quite well-known [5–9]. Studies of technological features of these processes revealed one of the most promising ways of processing charge materials using a highly concentrated heat source (plasma arc) that makes it is possible to produce welding materials for arc welding and submerged arc welding. The essence of this technology lies in the influence of a direct-action plasma arc on a charge of mineral raw materials in order to produce fused welding flux granules.

Therefore, the aim of this work is to identify characteristics of welding flux granules formation in carbon steel surfacing when exposed to direct action plasma arc on the charge of mineral raw materials from the Ural region.

## 2. Selection of raw materials and technical parameters of products

Rocks with oxide-silicate base from the Ural region gabbroid group were chosen as mineral raw material for observation. These raw materials contain all the necessary components for fused flux slag base and low content of harmful impurities. Hornblendite and other rocks of the Ural region gabbroid group have good fusibility, required melting point, homogeneity in chemical and phase composition and a number of other geological advantages [10, 11].

Possible production volume is a technical parameter that determines the quantitative characteristics of fluxes produced by electric arc method and it depends on equipment transmission capacity. Technical parameters that determine quality characteristics of the equipment and surfacing fluxes being developed are the following: structural elements and complexity of the equipment design, chemical composition of developed fluxes, structure and size of granules and bulk density. These parameters affect the quality of welded joints of critical structures produced by welding under a layer of developed flux.

## 3. Description of the experiments

Welding materials with a high deposition rate are mainly used for surfacing. That is achieved by adding metal powder to the electrode coating mixture during production of coated electrodes or to the charge during production of ceramic flux [12, 13]. Figure 1 shows granules produced by granulating of metal plate with a raw mineral substrate using a highly concentrated heat source (plasma arc).

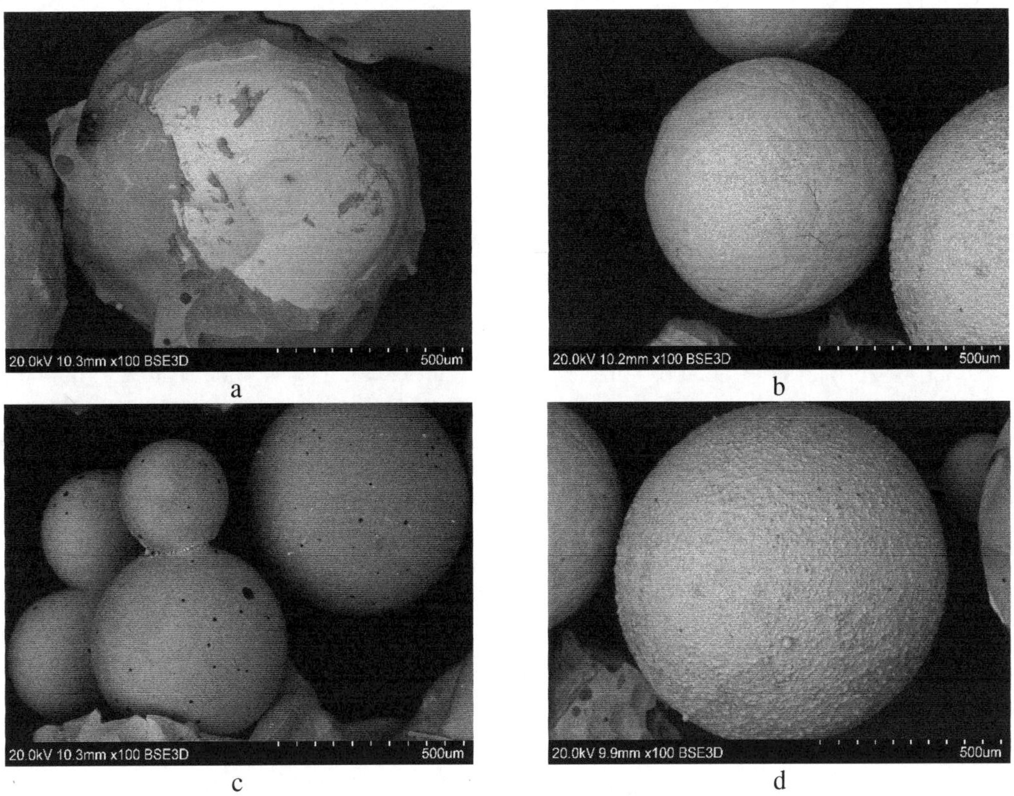


**Figure 1.** Formation of welding flux granules, different in morphology and chemical composition, x100: a – granule on metal base, coated partly by slag; b – granule coated with slag; c – porous granules on slag base; d – oxidized metal granule.

Under plasma arc influence ignited on metal plate made of Steel 20 (similar to S235) with a thickness of not more than 1 mm combined with a substrate 10-15 mm thick from finely dispersed mineral raw materials mixed on liquid sodium glass, there is a spraying of metal layer and subsequent melting of substrate that forms granules coated with an oxide-silicate base (figure 1, a, b). Scanning electron microscopy (electron microscope Carl Zeiss EVO50) was used to study the morphology and chemical composition of granules and the results of studies determining chemical composition of granules are presented in table 1.

**Table 1.** Results of chemical analysis of fused welding flux granules.

| Granules in Figure 1 | Chemical composition, wt. % | | | | | | | | | |
|---|---|---|---|---|---|---|---|---|---|---|
| | O | Na | Mg | Al | Si | Ca | Ti | Cr | Mn | Fe |
| a (slag) | 35.35 | 2.14 | 3.19 | 2.76 | 12.45 | 4.53 | 0.38 | 0.58 | 0.19 | 38.42 |
| a (metal) | 5.29 | 1.46 | 0.62 | 0.39 | 1.04 | 0.18 | - | - | - | 91.02 |
| b | 28.92 | 0.96 | 0.96 | 0.76 | 3.79 | 1.61 | 0.08 | 0.35 | 0.30 | 62.27 |
| c | 22.01 | 0.56 | 0.08 | 0.11 | 0.63 | 0.72 | - | - | 0.16 | 75.73 |

High content of sodium oxide is a result of using a binder of liquid sodium glass for mixing the oxide-silicate base of the substrate and bonding with metal plate. Magnesium, aluminum and other metal oxides are present in a small amount, since their content in mineral raw materials substrate does not exceed 7 wt. %, as well as in the process of granulation it is partially burned out under the influence of plasma arc. If slag crust was formed on the surface of the granules, iron content exceeded more than 90 wt. % in granule center. Otherwise, without the formation of slag on the surface of the granules, the iron was oxidized and its content reached 38.42 wt. % (figure 1, a). Studies have shown that volume of granules coated with slag reaches up to 95%.

The form of welding flux granules is determined by external and surface tension forces influence. The basis of the principle of spherical shape formation of granules lies in equilibrium state. When external forces are absent or compensated, the surface of slag melt tends to form a spherical shape to have a minimum area.

Without external factors influence, there is separation of the granule from the plate when its weight becomes equal to the sum of surface tension forces. Then the granule from the melted bridge formed between the plate and a drop of slag comes off and falls freely and the remaining bridge begins to change its shape. Therefore, appearance of each large granule with a size of up to 4 mm is accompanied by formation of another small granule (figure 2, a). The volume of these granules has significantly lower values than the first granule. The above principle of granular formation is confirmed by experiment where under the influence of electric arc (diameter of electrodes is 6-18 mm) at 100-350 A current [14], there is melting of material in the form of plate and detachment of granules shown in figure 2, b. The plates used in the experiment that were subjected to melting consist of finely dispersed hornblendite powder from the Ural region, mixed in a 2: 1 mass ratio with liquid glass.

In some cases, besides spherical granules, larger granules are formed due to the high viscosity of the formed slag, and also there are elongated fibers that are formed at the moment of detachment of granules from the plate when the bridge holds out to falling granule. It is possible to produce cooling on a rotating disk in order to avoid forming large granules with a fraction of more than 4 mm that are beyond the limits of regulatory documentation [15].

However, under dynamic influence of the plasma jet, process of granular formation occurs with forced detachment and acceleration of granules from the plate surface. In addition, granulation process using plasma arc instead of electric carbon one is characterized by the presence of a metal substrate that melts under plasma jet influence and gets in welding flux granules formation.

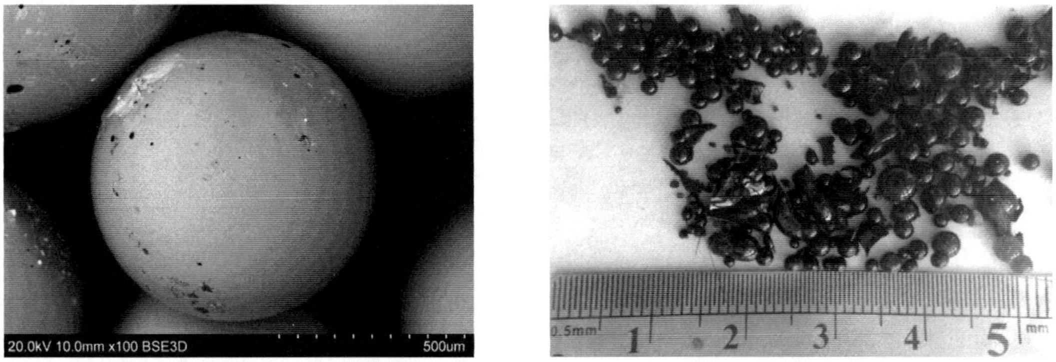


a b

**Figure 2.** Welding flux granules, produced under electric arc influence between carbon electrodes: a – spherical granules, x100; b – general view of granules produced by “dry” granulation.

It has been established that granulation with influence of plasma arc, in contrast to carbon electric one, there is a decrease in volume of formed agglomerates, i.e. chains of granules that are later melted into larger particles, with that, having a complex morphology. This result is achieved due to influence of dynamic effects of a highly concentrated heat source (plasma arc) that gives a high rate of falling of the melted granules into granulation pool with water that prevents the granules from interacting with each other. The lack of porosity of the granules is achieved because of their instant cooling due to the high falling rate and immediate solidification in granulation pool with water.

Preliminary studies of welding and technological properties of welding flux granules showed high performance rates of surfacing, but an adjustment of the charge composition is required for arcing stability and surfacing as a whole.

## 4. Conclusion

Developed fused welding flux can be used in production and repair works (surfacing) of critical structures made of carbon and low-alloyed steels of several groups of technical devices provided and classified by the National Agency for Welding Control.

Thus, a method for producing welding fused flux for surfacing using a highly concentrated heat source (plasma arc) has been proposed. With produced fused flux granules that contain both slag and metal base, it is possible to vary different properties such as composition and morphology of the granules within a wide range, producing welding materials of wide application.

## Acknowledgments

The reported study was funded by RFBR according to the research project № 18-33-00599.

## References

[1] Kartsev D S and Zernin E A 2016 Use of Refractory Nanoparticles as a Component of Welding Materials in Welding and Surfacing with Coated Electrodes and Flux Cored Wires *IOP Conference Series: Materials Science and Engineering* **142 (1)**

[2] Ignatov I M, Naumov S V and Ignatova A M 2013 *RU Patent 2494847* C1

[3] Gorynin I V et al. 2012 *RU Patent 2387521* C2

[4] Changming D, Chao S, Gong X, Ting W and Xiange W 2018 Plasma methods for metals recovery from metal-containing waste *Waste Management* **77** 373-38

[5] Acayanka E, Djowe A Tiya, Laminsi S, Tchoumkwé C C, Nzali S, Mbouopda A Poupi, Ndifon P T and Gaigneaux E M 2013 Plasma-assisted synthesis of $TiO_2$ nanorods by gliding arc discharge processing at atmospheric pressure for photocatalytic applications *Plasma Chemistry and Plasma Processing* **33 (4)** 725-35

IOP Conf. Series: Materials Science and Engineering **510** (2019) 012013 doi:10.1088/1757-899X/510/1/012013

[6] Djowe A T, Laminsi S, Njopwouo D, Acayanka E and Gaigneaux E M 2013 Surface modification of smectite clay induced by non-thermal gliding arc plasma at atmospheric pressure *Plasma Chemistry and Plasma Processing* **33** **(4)** 707-23

[7] Samal S 2018 Synthesis of $TiO_2$ Nanoparticles from Ilmenite Through the Mechanism of Vapor-Phase Reaction Process by Thermal Plasma Technology *Journal of Materials Engineering and Performance* **27** **(6)** 2622-28

[8] Swain B, Samal S K, Mohanty M K, Behera A and Mishra S C 2018 Study of effective utilization of iron ore sinter through arc plasma *IOP Conference Series: Materials Science and Engineering* **338** **(1)**

[9] Shanenkov I, Sivkov A, Ivashutenko A, Medvedeva T and Shchetinin I 2019 High-energy plasma dynamic synthesis of multiphase iron oxides containing $Fe_3O_4$ and $\varepsilon$-$Fe_2O_3$ with possibility of controlling their phase composition *Journal of Alloys and Compounds* **774** 637-45

[10] Artemov A, Ignatov M, Ignatova A and Naumov S 2017 Composition development and production technology of stone casting silicate materials and items *Key Engineering Materials* **743** 401-05

[11] Ignatov I M, Ignatova A M and Naumov S V 2015 Development of a slag base for welding materials from mineral and technogenic raw materials of the Urals *Ferrous Metallurgy. Bulletin of Scientific, Technical and Economic Information* **10** **(1390)** 67-79

[12] Ignatov I M, Ignatova A M and Naumov S V 2014 *RU Patent 2504465*

[13] Mikhailitsyn S V, Sheksheev M A, Platov S I, Emelyushin A N and Naumov S V 2018 Investigation of viscosity of liquid welding slags and melts of electrode coatings *Izvestiya Vysshikh Uchebnykh Zavedenij. Chernaya Metallurgiya* **61** **(4)** 280-87

[14] Naumov S V, Ignatov M N and Sheksheev M A 2017 Technology of mineral raw materials granulation by electric arc for manufacturing of welding fused flux *Solid State Phenomena* **265 SSP** 290-95

[15] Peng H, Shan X, Kang J, Ling X and Wang D 2018 Influence of rotary disk configurations on droplets characteristics in molten slag granulation for waste heat recovery *Applied Thermal Engineering* **135** 269-79

IOP Conf. Series: Materials Science and Engineering **510** (2019) 012014 doi:10.1088/1757-899X/510/1/012014

# Photo-induced effects in the doped ferroelectric

**Y Perkov, Yu Karpets and V Ivanov**

Department of Physics and Theoretical Mechanics, Far Eastern State Transport University, Khabarovsk, Russia

E-mail: tmeh@festu.khv.ru

**Abstract**. Light induced currents in thin lithium niobate crystal were described. New effects were observed in doped lithium niobate crystals with two electrodes of different metals (sandwich system). The light induced response includes the photovoltaic effect and the thermoinduced effect. Spectral and coordinate dependences of the light response are investigated.

## 1. Introduction

The photovoltaic effects are well known phenomena for the ferroelectric crystals [1-3]. Last time were discovered a series of nonequilibrium phenomena in ferroelectrics, which do not have a definite theoretical description [3,4]. In paper [5] the thermovoltaic effect consisting in initiation of the electric voltage between the opposite faces of the semiconductive sample of samarium sulphide (SmS) on heating to 400–500 K is described. The effect has appeared in the presence of the donor impurity concentration gradient directed towards these faces.

In this paper, the results obtained in experimental studies of the light induced currents in the thin-layer metal-lithium niobate-metal (MLM) structure are presented. The spectral and coordinate dependences of the light response are experimentally investigated. This effect is similar to photovoltaic effect, but it includes the thermal contribution. The possible models of the investigated phenomena are discussed. равна

## 2. Spectral dependence of the light induced response

The samples of doped lithium niobate crystals (iron concentration is equal to 0.3 wt%) with different pairs of electrodes were studied. The metal electrodes were applied by vacuum evaporation with thickness varying from 0.1 to 1 μm. The samples with different pairs of electrodes (Al – Cr, In – Cr, Al – Cu, Ag – Al) were used.

The study of the light induced effect was performed using the slow modulation of the light falling on the crystal. Because the sufficiently high rates of heating [about 0.1 K/s] were registered during the experiments, the pyroelectric current for the polar *Z*-cut crystals reached the considerable value. Therefore, the principal measurements were performed for the *Y*-cut lithium niobate crystals where the pyroelectric response could be observed in pure form.

It was experimentally measured the coefficient characterizing the light induced response:

$$G = \Delta U_n / P, (V/W), \quad (1)$$

where $\Delta U_н$ is the change of voltage across load impedance $R_n$ when the sample is illuminated by the light, P is the power of the incident radiation.

The measurement results are presented in Figure 1. It was used the $LiNbO_3$:Fe crystal (0.3 wt.% Fe, Al – Cr, $2 \times 2.5 \times 0.13$ $mm^3$). The preamplifier coefficient is K=20. The light response has a sharply defined spectral dependence with sign changing about 0.9 μm wave length.

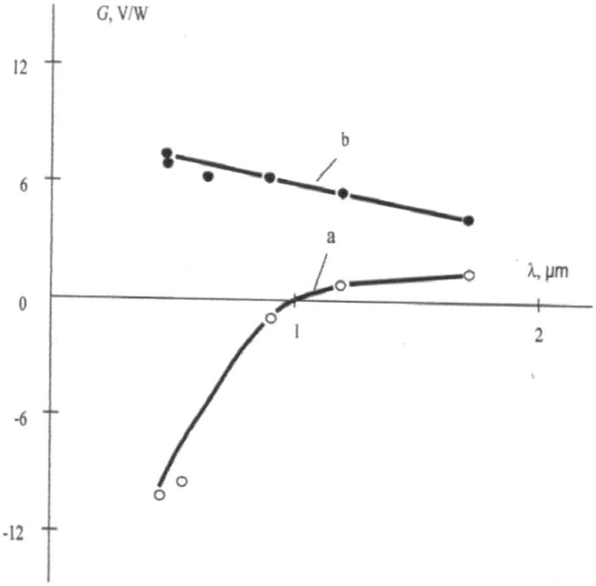


**Figure 1.** The photocurrent coefficient versus the radiation wavelength: a – the translucent electrode; b – the blackened absorbing electrode

The calibration measurements revealed a very weak (a few percent) spectral dependence of absorptive electrode. It can be concluded that the sharp spectral dependence of the light response is due to the bulk absorption in the crystal, because the electrodes are partially transparent. The presence of volumetric absorption suggests that changing of the signal character and its growth toward the short waves are induced by the fotogalvanic effect. The evaluation of Glass constants (in the field 0.5÷1.5 μm) is consistent with data available in the literature [3]. It can be assumed that the response in the infrared spectrum region is due to the thermovoltaic effect.

## 3. Thermovoltaic effect in the metal-ferroelectric-metal sandwich structure

In our work [6], the thermally stimulated quasi-stationary current arising only in the highly alloyed with iron crystal of lithium niobate with the evaporated electrodes from a pair of different metals was described. A sign of the thermoinduced electro-mowing force (EMF) is determined by location of electrodes applied by vacuum evaporation on the opposite faces of crystal and does not depend on orientation of the crystallographic axes of the sample in reference to electrodes.

In contrast with the classic pyroelectric effect, the value of observed thermal response is proportional to the crystal temperature increase and does not depend on the rate of its change. It was experimentally measured the coefficient characterizing the thermo-EMF value in the mode of voltage $U_n$ measurement:

$$P_{el} = (R_n S)^{-1} \Delta U_n / \Delta T, \text{ (A/K·cm}^2\text{)}, \quad (2)$$

where $S$ is the area of crystal surface covered with sputtered (evaporated) electrode; $\Delta U_н$ is the change of voltage across load impedance $R_n$ when the sample temperature changes by $\Delta T$; $\Delta T$ is the difference between the initial and final temperatures of sample [6].

The study of the effect was performed using the slow modulation of the crystal temperature under the conditions of thermostat and the thermo-EMF was recorded under the stationary conditions (relaxation was not observed for observation time $\tau \sim 10^5$ $s$). The measurements showed that the thermo-voltaic effect has approximately equivalent values in the $Y$- and $Z$-cut crystals. Because the sufficiently high rates of heating [about 0.1 K/s] were used in the installation, the pyroelectric current for the polar $Z$-cut crystals reached the considerable value. Therefore, the principal measurements were performed for the $Y$-cut lithium niobate crystals where the thermal response could be observed in pure form.

The metal electrodes were applied by vacuum evaporation with thickness varying from 0.1 to 1 μm. The samples with different pairs of electrodes were studied. For studying the dependence of the thermo-EMF value on the concentration of impurity, the lithium niobate crystals with the same areas and thicknesses [$S$ = 0.5 cm$^2$, $d$ = 1 mm] were used. The iron concentration varied in the range of $1.3 \cdot 10^{-2}$ wt% to 0.6 wt%. For the impurity concentration of less than 0.25 wt%, the thermo-voltaic effect is not observed. Beginning with the impurity concentration of 0.25 wt%, the coefficient $P_{el}$ rises sharply and reaches maximum ($P_{el} = 5 \cdot 10^{-12}$ A/cm$^2$K) at concentration of 0.3 ÷ 0.4 wt% of Fe.

When investigating the dependence of the coefficient $P_{el}$ value on the sample geometry, it was identified that $P_{el}$ value increases sharply when the crystal thickness decreases (Figure 2).

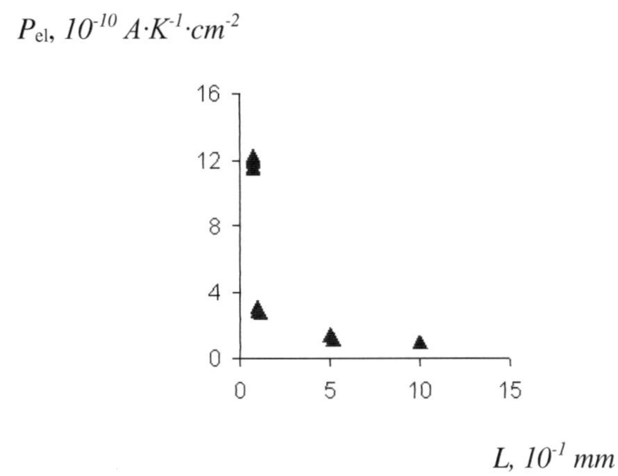


**Figure 2**. The coefficient $P_{el}$ versus the crystal thickness ($S$=5 $mm^2$; 0.3 wt% of $Fe$; Al–Cr electrodes.

The known contact phenomena in the sandwiched structures with ferroelectrics do not explain the observed quasi-stationary currents [7,8]. The presence of initial voltage $U_n^0$ across the sample allows us to suggest the non-equilibrium nature of the thermally stimulated current. In our work [9] it was proposed the electret model of the MLM structure. In this model, the thermoinduced changing of the conductivity of the crystal leads to change in the stationary current on the load impedance [10].

## 4. Coordinate dependences of the light induced response

In experiments using as the source of radiation He-Ne laser (0.63 μm) it was revealed the dependence of light induced response from the coordinates of the laser spot on the crystal top surface (Figure 3).

The crystal was installed in the coordinate table, which allows you to seamlessly move crystal in two directions. The light spot diameter in the focus lens was 340 μm.

IOP Conf. Series: Materials Science and Engineering **510** (2019) 012014 doi:10.1088/1757-899X/510/1/012014

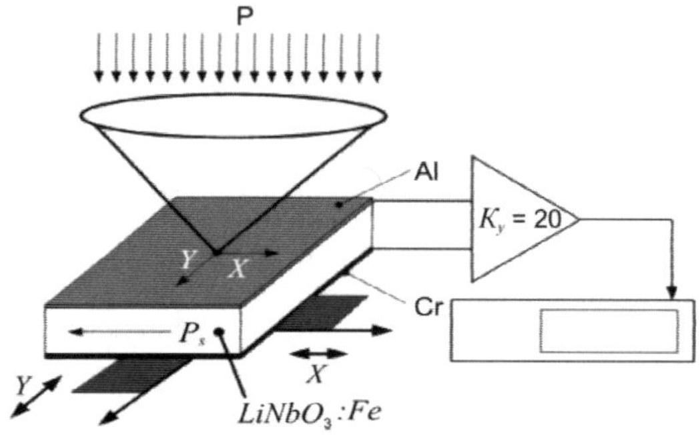


**Figure 3**. Experimental scheme.

The value of the stationary light response versus X (along polar axes) and Y-coordinates is shown in Figure 4a. The coefficient $G^*$(G in arb. un.) reaches the greatest magnitude at the ends of the crystal and it takes different signs at the opposite ends of the sample. This asymmetry can be associated with only one selected direction in crystal associated with the polar axis X. The coefficient $G^*$ versus Y-axis is changed monotonically (Figure 4b).

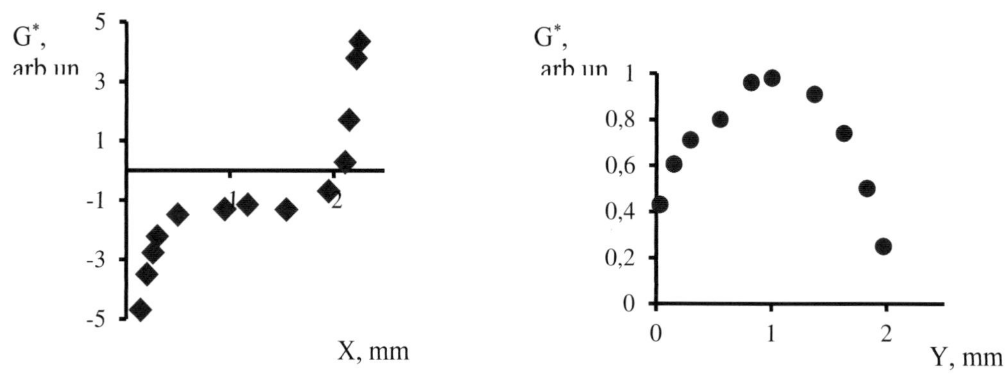


a) b)

**Figure 4**. The coefficient $G^*$ versus the coordinate X (a) and the coordinate Y (b) (lithium niobate crystal; 0.3 wt% of Fe; Al–Cr electrodes; 2 × 2.5 × 0.13 $mm^3$).

The anomaly in the coordinate dependence can be explained by the existing two response mechanisms (photovoltaic and thermovoltaic) with different signs. The photovoltaic response gives the contribution of light induced current, because the electrodes were partially transparent. This is confirmed by the next experimental fact. When the upper electrode was blackened this coordinate dependence исчезает disappears. The detail explanation of the describing experiments requires the next investigations.

## 5. Conclusions

The results obtained in experimental studies of the light induced currents in the thin-layer metal-lithium niobate-metal structure are presented. In the nonsymmetrical MLM system is registered the stationary light induced current. The light response has a sharply defined spectral dependence with sign changing. In experiments with the local illumination of the $LiNbO_3$:Fe crystal it was revealed the anomalous dependence of light induced response from the coordinates of the laser spot on the crystal top surface. The alleged explanation of the both effects includes the photovoltaic effect and the thermovoltaic contribution.

IOP Conf. Series: Materials Science and Engineering **510** (2019) 012014 doi:10.1088/1757-899X/510/1/012014

The results obtained can be used for investigating the properties of the sandwiched ferroelectric structures [11-13] as well as for developing new types of the photo-electrical energy converters [14-16].

**References**

[1] Serway R A 1990 *Physics for Scientists & Engineers* (3rd ed.) Saunders p 1150
[2] Glass A, Von der Linde D and Nergran T 1974 *Appl. Phys.* **25** 233
[3] Kuz'minov S 1990 *Ferroelectric Crystals for Laser Radiation Control* (Bristol :Adam Hilger).
[4] Volk T and Woehlecke M 2008 *Lithium Niobate: Defects, Photorefraction and Ferrolectric Switching* (New York: Springer)
[5] Kaminskii V 2013 *Technical Physics Letters* **39** 673
[6] Itskovskii M, Shchedrina L and Kladkevich M 1980 *Ferroelectrics* **29** 167
[7] Ivanov V I, Karpets Y M and Perkov Y O 2017 *Proc. SPIE* **10176** 1017625
[8] Pichugin V F, Bulycheva A A, Yakovlev V Yu and Kim I W 2005 *Physica status solidi (c)* **1** 208-211
[9] Barkan I B, Baskin E M and Entin M V 1980 *Phys. stat. sol. (a)* **59** 12-25
[10] Kim I W, Park B C, Jin B M, et al. 1995 *Material Letters* **24** 157-160
[11] Ivanov V I, Karpets Y M and Perkov Y O 2017 *IOP Conf. Ser.: Mater. Sci. Eng.* **168** 012034
[12] Romero J J, Jaque D., Garcia-Sole J. and Kaminskii A A 2001 *Appl. Phys.Lett.* **78** 1961
[13] Gladki V V, Ivanova E S and Volk T R 2007 *Phys. Solid State* **49** 2148
[14] Ivanov V I and Karpets Y M 2019 *Proc. SPIE* **11024** 110240D
[15] Pertsev N A, Petraru A, Kohlstedt H, Waser R,. Bdikin I K, Kiselev D and Kholkin A L 2008 *Nanotechnology* **19** 375703
[16] Perkov Y O and Ivanov V I 2018 *J. Phys. Conf. Ser.* **1038** 012114

IOP Conf. Series: Materials Science and Engineering **510** (2019) 012015 doi:10.1088/1757-899X/510/1/012015

# Simulation of thin-walled workpieces ends expanding for pipelines making

**D A Potianikhin, B N Maryn, S I Feoktistov and Kyaw Zayar Soe**
Komsomolsk-on-Amur State University, 27 Lenin Ave., Komsomolsk-on-Amur, 681013, Russia

Email: potyanikhin@mail.ru

**Abstract.** Experience of pipelines operation shows that their failures are often connected with thinning of walls in bending areas and with abrupt transition from one wall diameter to another. It is possible to make this transition smoother by using of connecting pipes with variable diameter. Such parts can be obtained by forming process of expanding. In this paper, the process of cold expanding of the end of thin-walled tubular workpiece, which is made from aluminium alloy AMg6M, against hard die, is investigated by finite-element method in ANSYS. It is used the saddle-shaped die, which makes it enable to produce connecting branches similar to the form of two joined pipes with specified blending radius. It has been investigated maximal and residual stresses in deformed part, as well as wall thinning. This stresses do not exceed tensile strength of the material. Therefore, it has been proved the possibility to obtain connecting pipes per one processing step, which do not require subsequent mechanical treatment. Wall thinning corresponds to technological norms.

## 1. Introduction

Metal forming is the process of parts manufacturing by force action on initial workpiece. This process is one of the main methods of parts manufacturing in mechanical engineering because of its relatively low cost and materials consumption, as well as relatively high efficiency.

Pipelines are one of the most important structural elements in mechanical engineering. They work under complex loading conditions. Pipelines are subjected to high pressure, pulsating loads and hydraulic shocks. Therefore, they have to meet the requirements of physical and mechanical properties of materials, surface finish, distortion of section form, and pipe wall thinning.

Working efficiency of pipelines can be reduced in consequence of excessive circular deviation and waviness of walls, that is, objectionable phenomena which accompany the process of workpiece deformation. Experience of operation shows that pipelines service failures are in most cases related to thinning of walls in bending points and with abrupt transition from one wall diameter to another [1]. One of the ways to make this transition smoother is using of connecting pipes, one of the ends of which has greater size than the other. Such parts can be obtained by forming process of expanding [2-8].

Defects of welding bead at the joints of the main and branch pipes can be one more significant problem during pipelines production. It may occur that the electrode metal penetrates into the pipeline and reduces the cross-sectional area of the branch pipe (Figure 1a). One of the ways to overcome this problem is also the creation of connecting pipes, in which one end is previously subjected to the forming process of expanding. In this case, the welding bead is moved to the zone of the header pipe, and even in the event of defect appearance, the area of its cross-section varies insignificantly (Figure 1b).

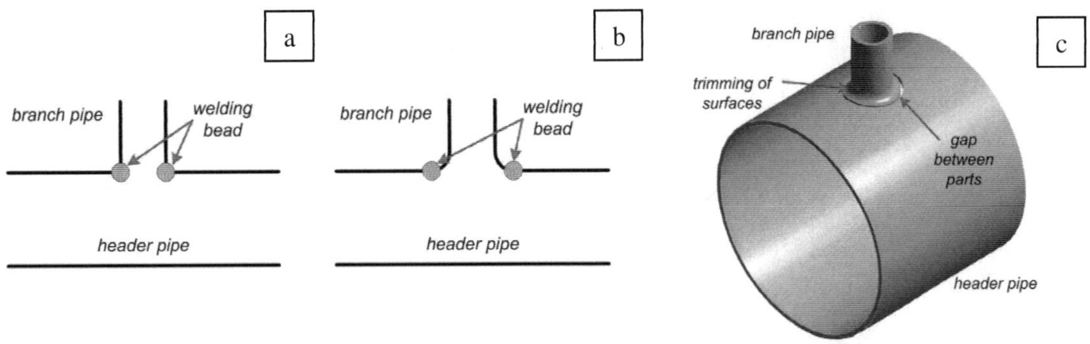


**Figure 1.** Probable defect of pipeline (reduction of cross-sectional area of a branch pipe) and method of its elimination: welding bead in the area of branch pipe (a) and in the area of header pipe (b); advisable shape of branch pipe (c).

However, it is necessary to take into account that weld width is approximately 1 mm. Therefore it is advisable that after expanding process the shape of branch pipe as much as possible replicates the form of two joined pipes with specified blending radius. With this purpose in mind, in present work the process of connecting pipe forming by expanding of cylindrical tubular workpiece is investigated by FEM simulation in ANSYS Workbench. It is used the saddle-shaped die, which shape makes it enable to produce connecting branches similar to ones in Figure 1c. Residual stresses and elastic strains, plastic strains boundary and part walls thinning are determined.

## 2. Geometrical model and finite element model

Geometrical model of the die and sectional view of tubular workpiece executed in ANSYS DesignModeler is shown in Figure 2a. Workpiece height $l$ = 30 mm, outer diameter $d$ = 8 mm, wall thickness $h$ = 0.8 mm. This shape of the die was obtained as union of circular cylinder, corresponding to inner volume of branch pipe, and a part of circular cylinder, corresponding to inner volume of header pipe. Axes of cylinders intersect orthogonally, and their surfaces are blended. Die dimensions are presented on section planes *OXY* and *OYZ* in Figures 2b and 2c. Blending radius $R$ is varied.

Finite element model is shown in Figure 4. Coordinate planes *OXY* and *OYZ* are the planes of symmetry, therefore, with the purpose of problem size reduction, calculations have been carried out for a quarter of geometrical model. Die is considered as perfectly rigid body, and so only its surface contacting with workpiece is of interest. Finite element model of the die contains one element by thickness. Its boundary conditions correspond to rigid attachment.

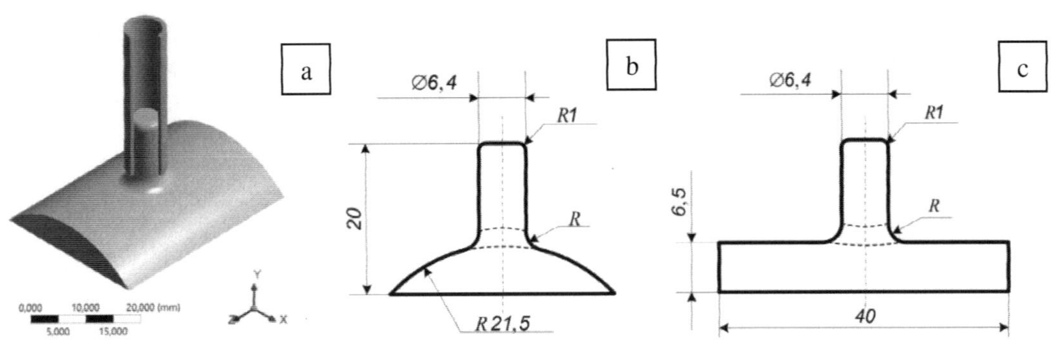


**Figure 2.** Geometrical model of tubular workpiece (in section) and die (a); sections of the die by planes *OXY* (b) and *OYZ* (c).

IOP Conf. Series: Materials Science and Engineering **510** (2019) 012015 doi:10.1088/1757-899X/510/1/012015

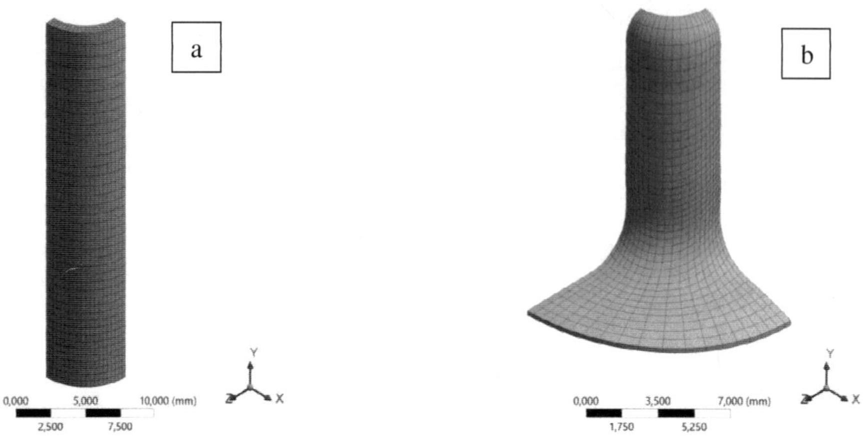


**Figure 3.** Finite-element model of tubular workpiece (a) and die (b).

It should be noted that previously mentioned assumptions has no effect on the results accuracy. For comparison purposes, it has been solved the problem where the die was considered as elastic-plastic solid presented in Figure 3, which is made from steel. Boundary conditions consisted in prohibition of vertical displacement for lower surface. Ultimately, deformations of the die proved to be negligible quantity. That is why it was decided not to take them into account for calculations time reducing.

After discretization the die is composed of 1200 higher order 3-D 20-node solid elements SOLID186. Tube wall is formed of 4000 elements SOLID186. It has been taken 4 elements by thickness. Coefficient of dry friction on contact surfaces is taken equal to 0.1. Contact and target surfaces are represented by elements CONTA174 and CONTA170, respectively.

Upper end of the tube is subjected to downward vertical displacement. Displacement value depends on blending radius: $s=R\pi/2$.

The problem has been solved in ANSYS Static Structural module. Stress-strain state is described with the use of elastoplastic model with linear isotropic hardening. Total strains are divided into reversible and irreversible components in accordance with relationship

$$\varepsilon_{ij} = \varepsilon_{ij}^{e} + \varepsilon_{ij}^{p} .$$

Elastic strains are connected with stresses by generalized Hooke's Law

$$\sigma_{ij} = \frac{E}{1+\nu}\varepsilon_{ij}^{e} + \frac{E\nu}{(1+\nu)(1-2\nu)}\varepsilon_{kk}^{e}\delta_{ij} ,$$

where $\sigma_{ij}$ and $\varepsilon_{ij}^{e}$ are components of stress tensor and elastic strain tensor, $E$ is Young's modulus, $\nu$ is Poisson's ratio, and $\delta_{ij}$ is Kronecker delta. The von Mises yield criterion has been used for irreversible deformations description. It can be written in terms of the principal stresses as follows:

$$(\sigma_1 - \sigma_2)^2 + (\sigma_2 - \sigma_3)^2 + (\sigma_3 - \sigma_1)^2 = \sigma_T^2 .$$

Here $\sigma_T$ is the tensile yield stress.

The tube is considered to be made from aluminium alloy AMg6M. Its physical-mechanical properties are given in Table 1.

**Table 1.** Physical-mechanical properties of aluminium alloy AMg6M [10].

| Parameter | Notation | Value | Unit of measure |
|---|---|---|---|
| Young's modulus | $E$ | $7.1 \cdot 10^4$ | MPa |
| Poisson's ratio | $\nu$ | 0.33 | |
| Tensile yield stress | $\sigma_T$ | 170 | MPa |
| Hardening modulus | $E_T$ | 71 | MPa |
| Ultimate tensile strength | $\sigma_B$ | 340 | MPa |

## 3. Results and discussions

The calculations have been made for four different blending radii from 1.5 mm to 3 mm in increments of 0.5 mm. In Figure 4, there are contour plots of von Mises equivalent stress $\sigma_{\text{eqv}}$ distributions at the moments of maximum vertical displacement of the tube:

$$\sigma_{\text{eqv}} = \left( \frac{(\sigma_1 - \sigma_2)^2 + (\sigma_2 - \sigma_3)^2 + (\sigma_3 - \sigma_1)^2}{2} \right)^{1/2}.$$

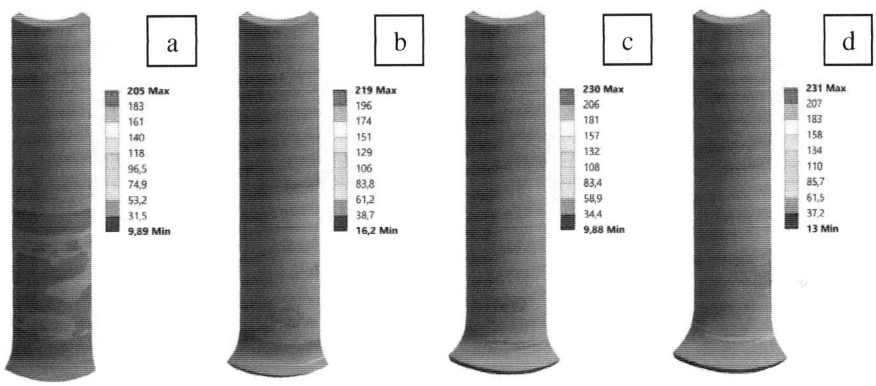


**Figure 4.** Equivalent von Mises stress distribution at the moment of maximal vertical displacement in deformed parts with blending radii $R$=1.5 mm (a), $R$=2.0 mm (b), $R$=2.5 mm (c), $R$=3.0 mm (d).

Residual stresses for the same geometrical parameters, obtained after removing of tool, are shown in Figure 5. One can see from the Figures 4 and 5 that stresses $\sigma_{\text{eqv}}$ do not exceed tensile strength $\sigma_B$. This means that mechanical strength of the part is ensured.

Figure 6 illustrates residual equivalent von Mises plastic strain

$$\varepsilon^p_{\text{eqv}} = \frac{1}{1+\nu} \left( \frac{(\varepsilon^p_1 - \varepsilon^p_2)^2 + (\varepsilon^p_2 - \varepsilon^p_3)^2 + (\varepsilon^p_3 - \varepsilon^p_1)^2}{2} \right)^{1/2}.$$

Hence, it can be determined the boundary of plastic deformations.

It has been investigated tube wall thinning. In Figure 7 it is demonstrated the thinning of bottom 7.5 mm length of deformed tubular part. Blue, red and green lines correspond to wall thinning in *OYZ* plane, *OXY* plane, and at the angle 45° between them, respectively. Maximal thinning does not exceed 32%.

IOP Conf. Series: Materials Science and Engineering **510** (2019) 012015 doi:10.1088/1757-899X/510/1/012015

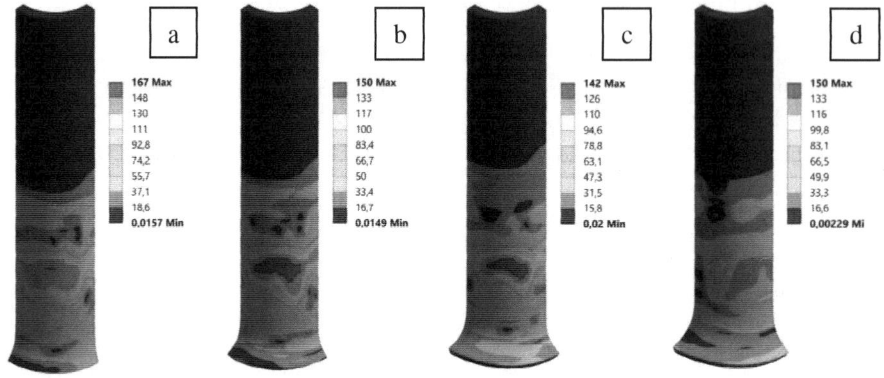


**Figure 5.** Residual equivalent von Mises stress distribution after pressing tool removing in deformed parts with blending radii $R$=1.5 mm (a), $R$=2.0 mm (b), $R$=2.5 mm (c), $R$=3.0 mm (d).

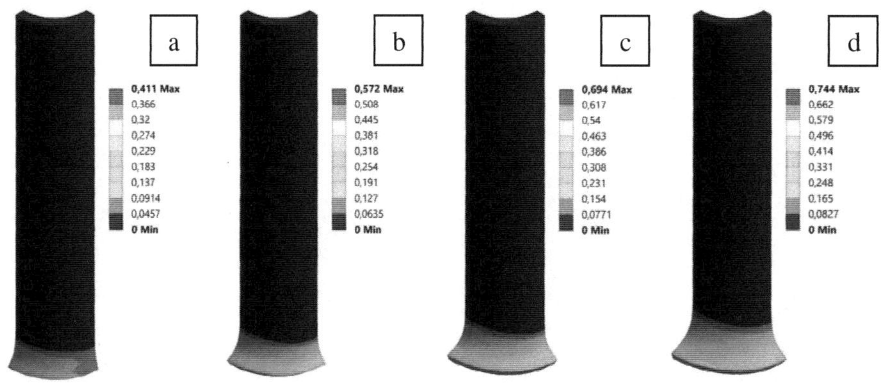


**Figure 6.** Residual equivalent von Mises plastic strain distribution after pressing tool removing in deformed parts with blending radii $R$=1.5 mm (a), $R$=2.0 mm (b), $R$=2.5 mm (c), $R$=3.0 mm (d).

## 4. Conclusion

Carried out calculations and the results of simulation show that it is possible to produce connecting pipes from tubular workpieces, which are made from aluminium alloy AMg6M, by compressing a thin-walled tube against saddle-shaped die. Stresses in deformed part do not exceed tensile strength. The gap between connecting pipe and the wall of header pipe is less than 0.5 mm and therefore it can be spanned during welding. At the same time, if mentioned above defect occurs it cannot significantly reduce cross-sectional area of a branch pipe. Thus, it has been proved the possibility to obtain connecting pipes per one processing step, which do not require subsequent mechanical treatment.

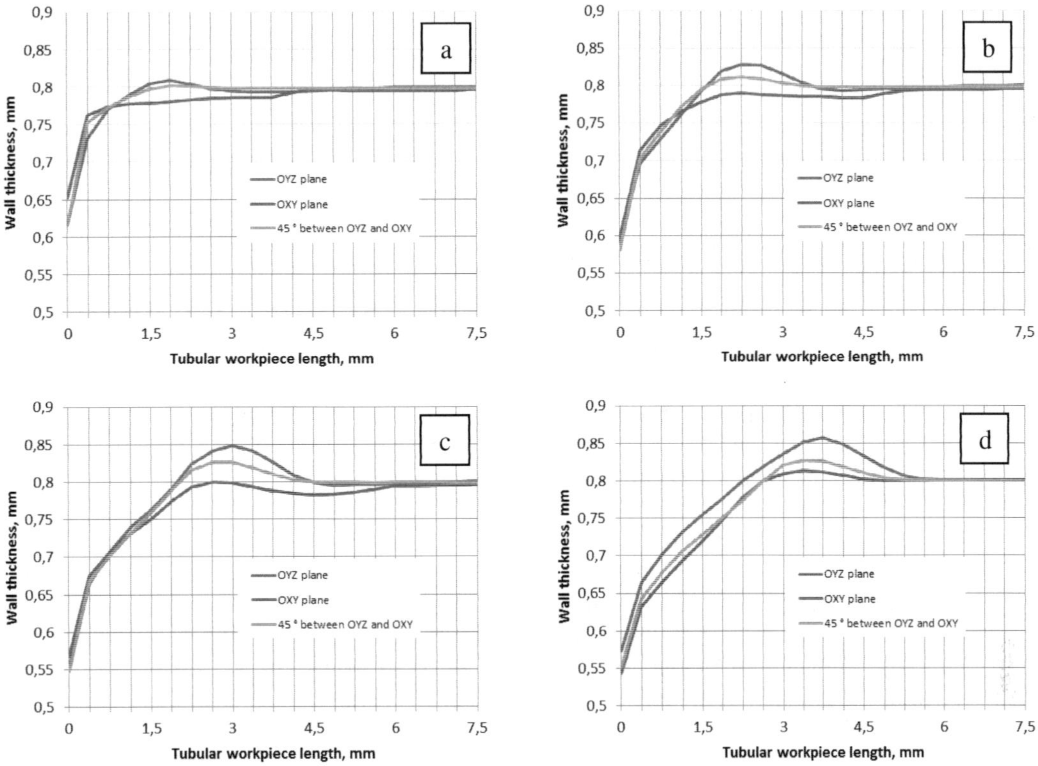


**Figure 7.** Wall thickness of deformed parts with blending radii $R$=1.5 mm (a), $R$=2.0 mm (b), $R$=2.5 mm (c), $R$=3.0 mm (d).

## References

[1] Maryn B N, Kim V A, Sysoyev O Ye, Kyaw Phone Htet and Hlaing Min Ko 2017 Defect analysis of operating hydrogasified piping systems *Uchenye zapiski Komsomol'skogo-na-Amure gosudarstvennogo tehnicheskogo universiteta* **1(29)** 79-89 [in Russian]

[2] Veselov A A 2012 Determination of the sizes of the end of pipes after the reconciliation by the method of plastic deformation *Morskoy Vestnik* **1** 15-16 [in Russian]

[3] Nepershin R I Thin-walled pipe expansion by a punch with a curvilinear profile 2010 *Journal of Machinery Manufacture and Reliability* **39 1** 66-72

[4] Almeida B P P, Alves M L, Rosa P A R, Brito A G and Martins P A F 2006 Expansion and reduction of thin-walled tubes using a die: Experimental and theoretical investigation *International Journal of Machine Tools and Manufacture* **46** 1643–1652

[5] Alves M L, Almeida B P P, Rosa P A R and Martins P A F 2006. End forming of thin-walled tubes *Journal of Materials Processing Technology* **177** 183–187

[6] Karrech A and Seibi A 2010 Analytical model for the expansion of tubes under tension *Journal of Materials Processing Technology* **210** 356–362

[7] Schaeffer L and Brito A M G 2007 FEM Numerical Simulation and Experimental Investigation on End-Forming of Thin-Walled Tubes Using a Die *Steel Research International* **78** 798–803

[8] Sun Z C and Yang H. 2007 Study on forming limit and feasibility of tube axial compressive process *Journal of Materials Processing Technology* **187-188** 292–295

[9] Daxner T, Rammerstorfer F G and Fischer F D 2005. Instability phenomena during the conical expansion of circular cylindrical shells *Computer Methods in Applied Mechanics and Engineering* **194** 2591–2603

[10] Alieva S G, Altman M B, Ambartsumyan S M et al. 1984 *Industrial aluminum alloys: Reference book* (Moscow: Metallurgiya) p 527 [in Russian]

# Changing magnetic parameters of structural 30KH13 (30X13) steel in relation to the heat treatment and low-cycle loading

**V V Probotyuk, V F Novikov, K R Muratov and R A Sokolov**
Physics Department of Monitoring and Diagnostic Techniques, Tyumen Industrial University, Tyumen 625000, Russian Federation

E-mail: falcon.rs@mail.ru

**Abstract.** The paper considers the magnetoelastic demagnetization of low carbon steel 30X13 under the action of cyclic loading. The change of the piezo-magnetic effect on the state of steel and the number of loading cycles is studied. The study used a wavelet transform to describe the state of the material under cyclic loading. It is established that the wavelet transform reflects the processes of fatigue changes going in steel.

## 1. Introduction

The paper describes the results of the tests high-chromium 30KH13 (30X13) stainless steel containing 12-14% chromium. It is a high-strength hard steel. After hardening, it is used for manufacturing very solid details. An absence of other alloying elements and optimal chromium content provide this relatively low-cost material with anticorrosive properties. The steel is ferromagnetic, thus enabling the application of magnetic methods to monitor its quality and condition.

Various heterogeneities present in the steel structure exert a significant influence on the steel parameters, including its magnetic characteristics. Admixtures, dislocations and inclusions mostly affect the formation of magnetic domains and their structures, influencing the magnetic characteristics of the material [1-3]. Mechanical varying loads and resulting flaws lead to irreversible reductions of the ferromagnetic remanent magnetization due to the domain structure transformation and domain wall motion [1, 3].

The work [4] analyzes magnetoelastic demagnetization of low-carbon steel under cyclic loading. It states that the magnetization peak values are approximated by the power function with parameters depending on the load amplitude. The work considers the possibility to estimate the steel fatigue limit, judging upon the approximation coefficients behavior. The work [4] also shows that 30KH13 (30X13) steel has the property of magnetoelastic power relaxation. This approach, however, takes into account the values corresponding only to the minimal and maximal loads, in other words, the magnetic cyclograms do not consider intermediate information.

## 2. Experimental procedures

This work aims at investigating the influence of cyclic loading of 30KH13 (30X13) steel on changing its magnetoelastic signal through wavelet analysis.

The analysis involved magnetoelastic cyclograms obtained in the work [4]. These were magnetoelastic cyclograms of 30KH13 (30X13) steel samples quenched and tempered at various temperatures (200°C-

650°C). Figure 1 presents magnetoelastic cyclograms for a 30KH13 (30X13) steel sample tempered at 530°C and loads of 160 and 960 MPa.

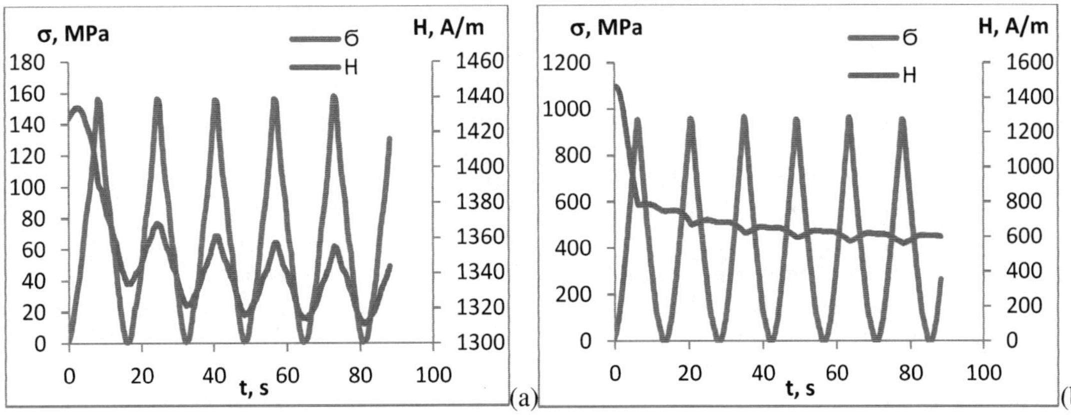


**Figure 1.** Time dependence of magnetic field intensity and changing mechanical stresses for a sample subjected to the tempering temperature of 530°C at the mechanical load amplitude of: $\sigma$=150 MPa (a), and $\sigma$=960 MPa (b).

The experimental curves obtained made it possible to reveal that, for the steel mentioned, within the given treatment temperature range and beginning with 450°C, the sign of the magnetic piezo effect changes with the mechanical load increase and this effect continues at higher temperatures. This phenomenon is not observed at lower temperatures. Figure 2 presents the results of change in the magnetic piezo effect at treatment temperature 530°C.

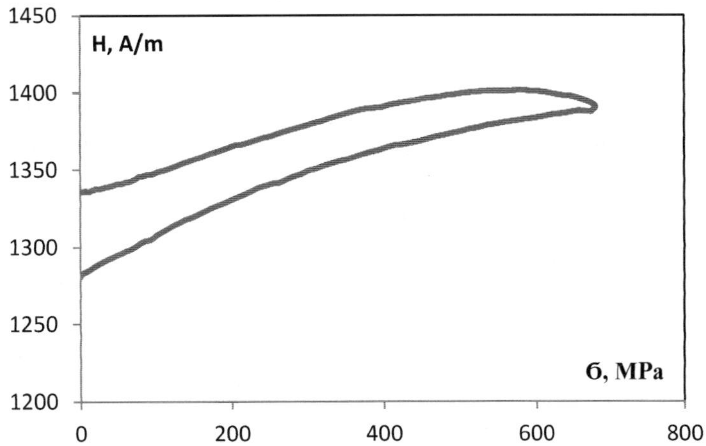


**Figure 2.** Load dependence of magnetic field intensity for a 30KH13 (30X13) steel sample at the tempering temperature of 530°C in the 40th loading cycle.

The direct piezo magnetic effect means the correlation between the minimum and maximum points of the maximal magnetic field intensity [5-6] and the load minimum and maximum points in the cycle. The reverse piezo magnetic effect implies an antiphase correlation between the values of magnetic field intensity and cyclic load.

## 3. Results and discussion

Figure 2 shows that the piezo magnetic effect transforms from the direct to the reverse one at 550 MPa. With the increase of the cycle number, this transformation shifts to the lower load range. Changing the sign of the piezo magnetic effect may be due to some more intense changes in the steel structure. These changes result from the coarsening of cementite particles due to the coalescence (the cementite coagulation). The steel structure is formed by the ferrite and cementite phases, also called sorbite [7-8]. These cementite particles serve as barriers for domain wall motion and rotation of magnetization vector. External loads constitute an additional energy source, helping the domain walls overcome structural heterogeneities [3, 7-8, 10].

In this work it is proposed to analyze the magnetic cyclogram in relation to the tempering temperature via wavelet analysis, which enables more detailed presentation of the data obtained.

An algorithm was developed in Mathcad CAS (Computer Algebra System) to investigate the influence of the cyclic load via unidimensional and two-dimensional analyses on the basis of the first derivative of the Gaussian function. The correlation was analyzed between a wavelet response of the magnetoelastic signal at low-cycle loads and mechanical properties of 30KH13 (30X13) steel (figure 3).

The obtained magnetic cyclograms, which were processed through unidimensional discrete wavelet transform, allowed to determine coefficients, and their values make it possible to assess the processes the steel undergoes (figure 3, 4).

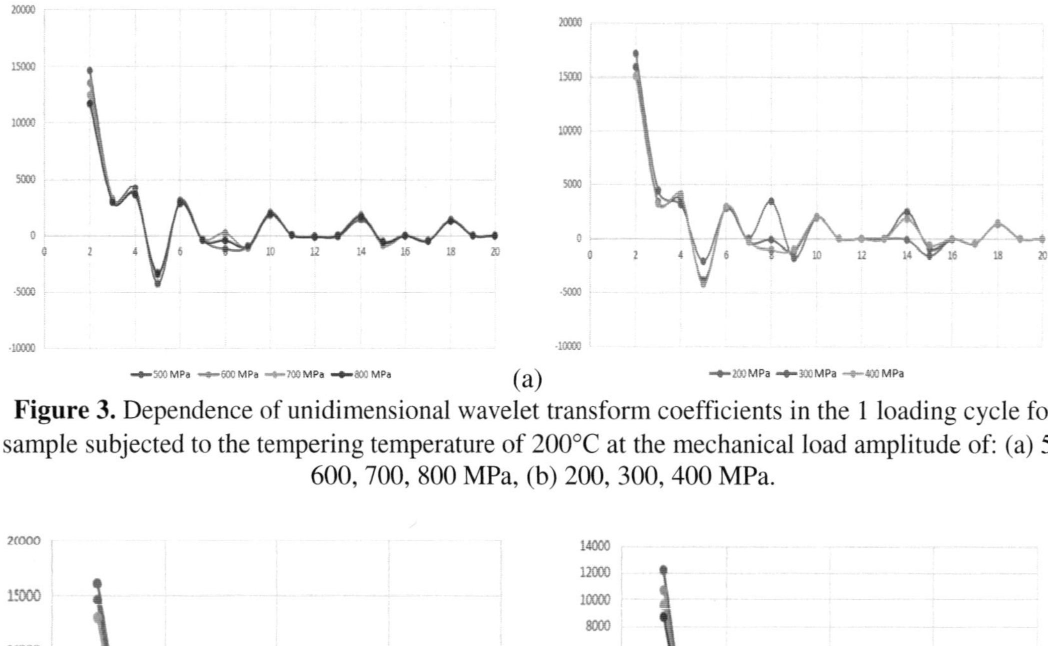


(a) (b)

**Figure 3.** Dependence of unidimensional wavelet transform coefficients in the 1 loading cycle for a sample subjected to the tempering temperature of 200°C at the mechanical load amplitude of: (a) 500, 600, 700, 800 MPa, (b) 200, 300, 400 MPa.

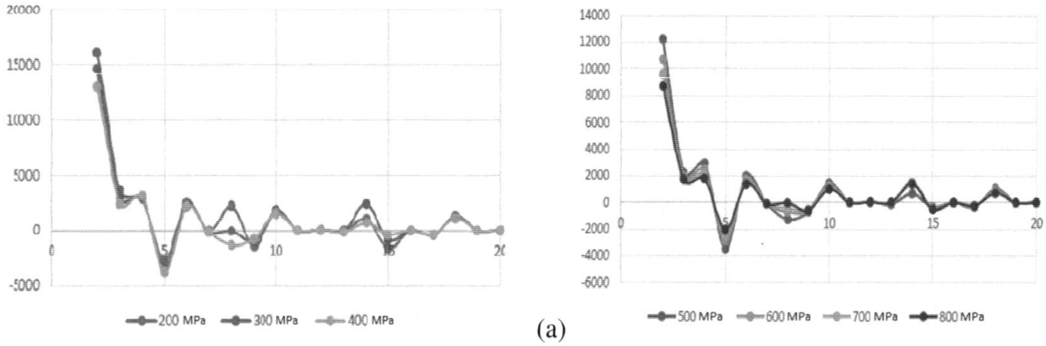


(a) (b)

**Figure 4.** Dependence of unidimensional wavelet transform coefficients in the 99 loading cycle for a sample subjected to the tempering temperature of 200°C at the mechanical load amplitude of: a) 200, 300, 400 MPa, b) 500, 600, 700, 800 MPa.

IOP Conf. Series: Materials Science and Engineering **510** (2019) 012016 doi:10.1088/1757-899X/510/1/012016

Having analyzed all the changes in the values of the coefficients obtained (figure 3, 4), we can conclude that, with the increase of the cycle number, as well as with the load increase, the values of unidimensional wavelet transform coefficients decrease, depending apparently on the structural heterogeneities. In addition, with the increase of the cycle number, the values of the coefficients align with the load increase.

Apart from unidimensional wavelet transform, the obtained magnetic cyclograms were analyzed via two-dimensional discrete wavelet transform (figure 6, 7). The data obtained were presented as horizontal relief maps; the graphs depict Z-axis as a level lines (see figure 5), differently colored, where the "red area" refers to the maximum coincidence of the magnetogram with the basic wavelet and the "blue area" refers to the minimum.

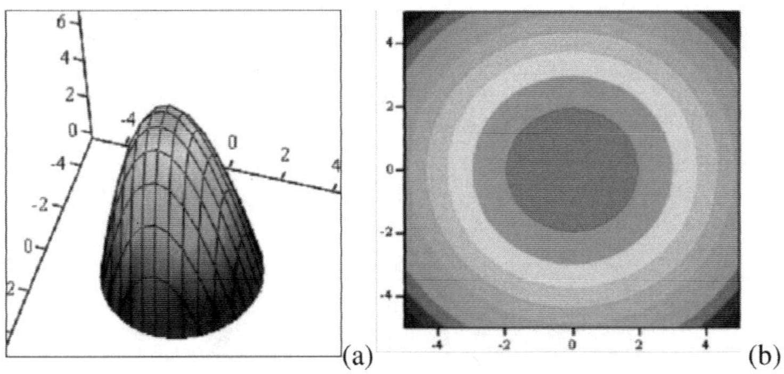


**Figure 5.** Data presented as a 3D graph (a) and as horizontal relief maps (b).

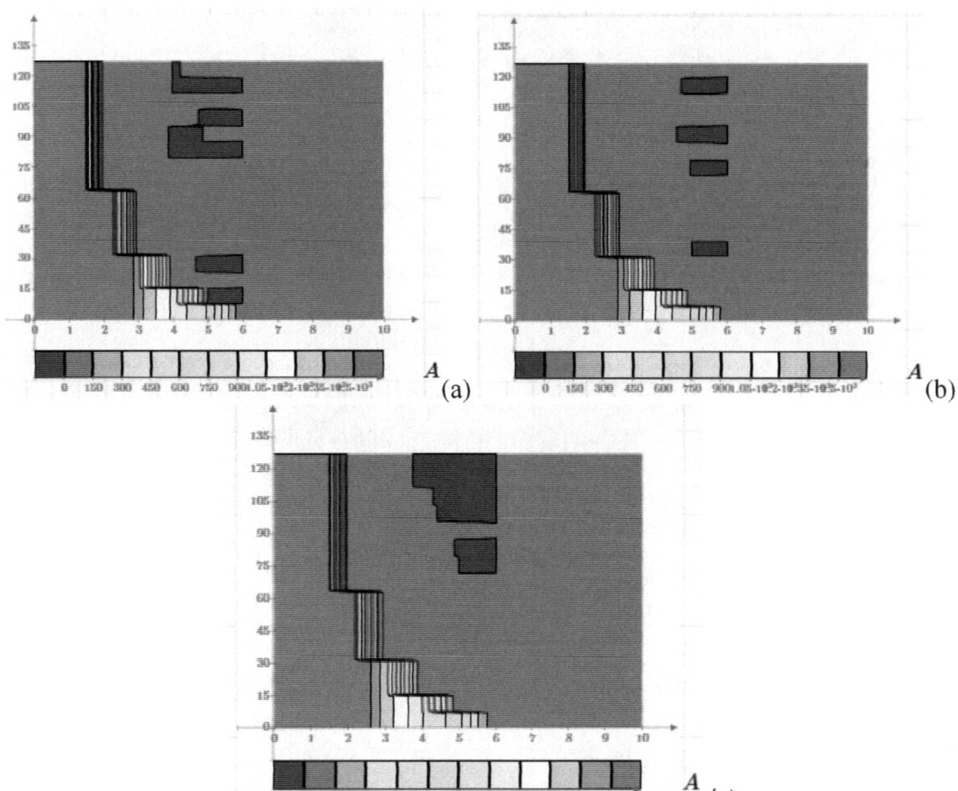

**Figure 6.** Dependence of two-dimensional wavelet transform coefficients in the 1 loading cycle for a sample subjected to the tempering temperature of 500°C at the mechanical load amplitude of: a) 200 MPa, b) 500 MPa, c) 700 MPa.

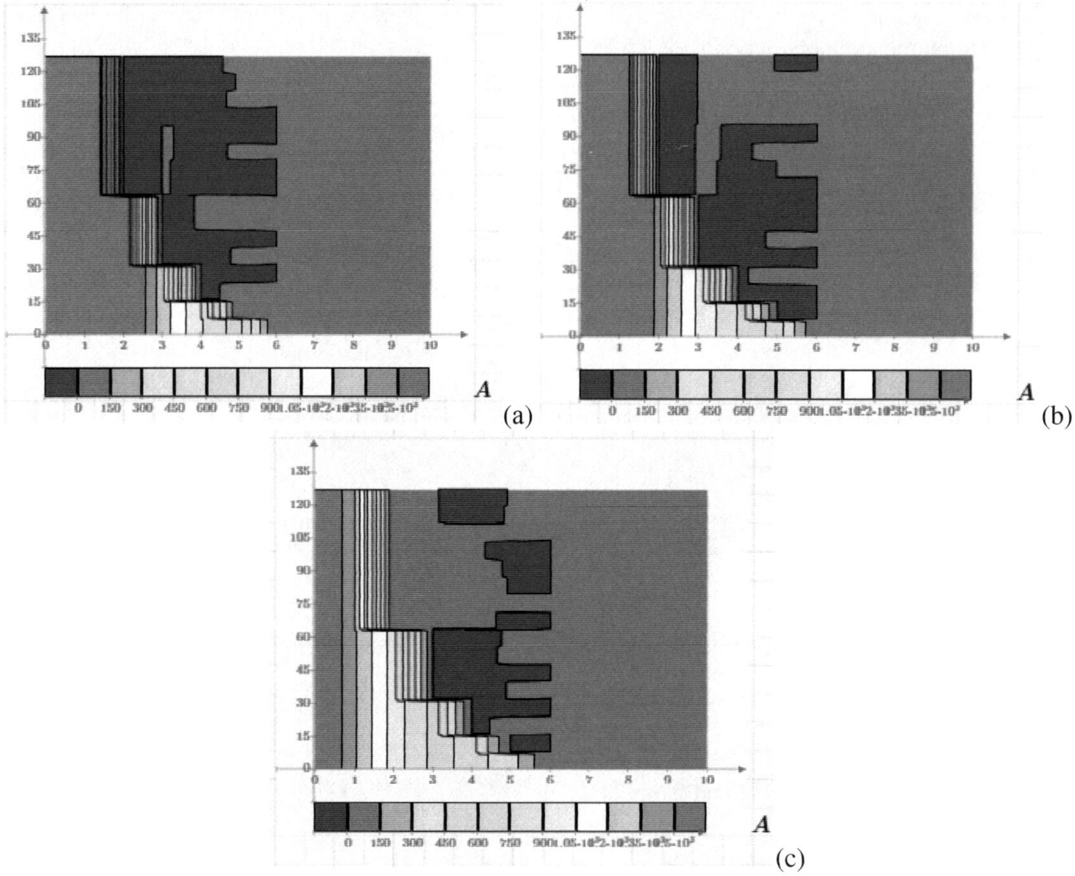


**Figure 7.** Dependence of two-dimensional wavelet transform coefficients in the 99 loading cycle for a sample subjected to the tempering temperature of 500°C at the mechanical load amplitude of: a) 200 MPa, b) 500 MPa, c) 700 MPa.

Figures 6 and 7 show that the “red area” is the largest at the initial load of 200 MPa and decreases monotonically with the increase of the cyclic load amplitude.

We also may suppose that the increase of the “red area” for the samples subjected to the tempering temperature range 200÷500°C is connected to intense structural changes, namely, transformation of the martensite structure into the pearlite one. This behavior is supposed to reflect the mechanisms of structural stress relaxation without microplastic deformations. The martensite structure is known to be stronger than the pearlite one. Thus, microplastic deformations of crystallites may be supposed to take place.

## 4. Conclusions

Two-dimensional wavelet transform analysis on the base of WAVE wavelet proved the following regularity: with the increase of the cyclic load amplitude, the “red area” (area with red color on the figure 6 and 7) decreases monotonically. The largest “red area” of two-dimensional wavelet transform is observed at the load of 550 MPa for all loading cycles, which may be connected to the structural features.

For 30KH13 (30X13) steel beginning with tempering temperature of 450°C, the sign of the magnetic piezo effect changes with the mechanical load increase.

## References

[1] Bida G V and Nichipuruk A P 2005 *Magnetic properties of heat-treated steels* (Yekaterinburg: The Institute of Metal Physics of the Ural Branch of the Russian Academy of Sciences) p 214

[2] Novikov V F and Bakharev M S 2001 *Magnetic diagnostics of mechanical stresses in ferromagnetics* (Tyumen: Vektor Buk Press) p 220

[3] Borovik Y S, Yeryomenko V V, Milner A S 2005 *Lectures on magnetism (3rd edition)* (Moscow: FIZMATLIT Press) p 512.

[4] Muratov K R, Novikov V F, Sokolov R A and Probotujk V V 2018 Influence of tempering temperatures on magnetoelastic relaxation behavior of 30KH13 steel *Key Engineering Materials* **769** pp 60-65

[5] Filatov V V 2008 *Magnetic prospecting: magnetic susceptibility, induced and natural remanent magnetization: scientific edition* (Yekaterinburg: The Ural State Mining University Press) p 236

[6] Novikov V F, Dyagilev V F, Bakharev M S, Nassonov V V and Prilutskiy V V 2006 Mechanisms of magnetoelastic change of local remanent magnetization for steels *Factory laboratory. Material diagnostics* **72** (6) pp 34-37

[7] Makovetskii A N and Mirzaev D A 2014 Influence of the interstitial range The Physics of *Metals and Metallography* **115** (6) pp 617-624

[8] Dragoshanskiy, Y.N., Pudov, V.I. 2016 Influence of structural heterogeneities on dynamics of magnetic domains and electromagnetic properties of fe-3%si alloy *Herald of Tambov University. Natural and technical sciences* **21** (3) pp 970-973

[9] Gorkunov E S and Ulyanov A I 1996 *Magnetic Methods and Tools for Testing the Quality of Powder-metallurgy Products* (Yekaterinburg: The Ural Branch of the Russian Academy of Sciences)

# The influence of PAR irradiance on yield growth of Chlorella microalgae

**O A Trofimchuk, P V Petikar, S B Turanov and S A Romanenko**

Division for Material Science, School of Advanced Manufacturing Technologies, National Research Tomsk Polytechnic University, 30 Lenin Avenue, Tomsk 634050, Russia

E-mail: oat1@tpu.ru

**Abstract.** The experiment indicated the influence of different PAR irradiance conditions on biomass growth of Chlorella vulgaris Beijer stock culture. The absorption spectrum of suspension was evaluated with spectrophotometer SF-256 UVI LOMO Photonika. In this work, effects of LED lamps with various radiant powers on yield gain were compared by analyzing the relation between cells concentration increase and time. It was concluded, that light is the most important factor influencing growth and cells development, as well as nutrient solution, temperature and carbon dioxide provision.

## 1. Introduction

Over the last decade, microalgae had been introduced to livestock animals' diet along with premix, vitamins and biologically active additives. Chlorella is a green microscopic algae species, widely regarded for its active biomass production [1]. It contains high-grade proteins, fats, carbohydrates and vitamins and is included in the category of "superfoods". Among all plants, Chlorella is at the top position in terms of many factors. For example, 40-60% of proteins, 30-35% of carbohydrates, 5-10% of lipids and up to 10% of mineral substances are included in the biomass of Chlorella [2].

Cell fission speed, which determines the survival of microalgae population, is mostly dependent on accessibility of energy sources and nutritional substances. As for all phototrophic organisms, light supply plays vital part throughout vegetation period of microalgae, while its properties, such as intensity and spectral composition, are of great importance [3, 4].

## 2. Materials and methods

Duration of the experiment – 54 hours.

Initial optical density of three samples: from 0.16 to 0.19 RU.

**Table 1.** Initial data.

| | *Installed power (W)* | *Ee(PAR) ($W/m^2$)* | *PPFD(PAR) ($\mu mol/s/m^2$)* | *PAR irradiance (nm)* |
|---|---|---|---|---|
| **Sample 1** | 4 | 10.3 | 48.6 | |
| **Sample 2** | 10.5 | 49.1 | 227.4 | 400…700 |
| **Sample 3** | 15 | 67.9 | 321 | |

IOP Conf. Series: Materials Science and Engineering **510** (2019) 012017 doi:10.1088/1757-899X/510/1/012017

Temperature (T, $^0$C): during the entire experiment temperature was maintained at optimal range 36.5 – 37.5 $^0$ C.

The experiments were carried out to observe the influence of irradiating source power on Chlorella microalgae cells concentration growth. Method was developed for the current labor facility, which was used for conducting the experiment in cultivating microalgae Chlorella.

Measurements of optical density spectrum were performed throughout the entire experiment period once every 3 hours 3 times a day with spectrophotometer SF-256 UVI. By analyzing concentration-time ration, we aimed to evaluate, how radiation, provided by LED lamps with diverse output, affected yield gain of the culture.

The schematic diagram of experimental facility is illustrated in figure 1. The setup consists of a parallelepiped-shaped case (2) and cover plate (1) made of Plexiglas [5].

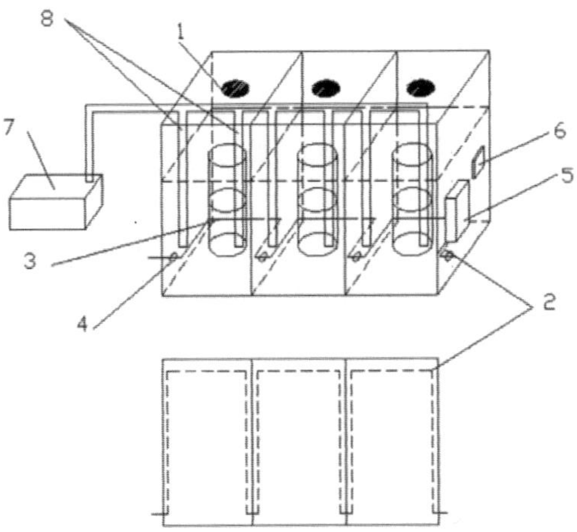


**Figure 1.** Scheme of experimental setup, (a) top view, 1 – cover plate; 2 – case; 3 – irradiating source; 4 – heating element; 5 – measuring containers; 6 – ports for heating element; 7 – thermostatic regulator; 8 – deferential breaker; 9 – pipes of aerator; 10 – aerator; 11 – ports between units.

300 ml measuring containers (5) are installed at the center of each unit of photobioreactor for the direct cultivation of Chlorella. Containers are filled with 250 ml suspension, contained 50% Tamiya medium and Chlorella stock culture.

To ensure temperature conditions, the case of photobioreactor (1) is flooded with water. Water temperature is maintained by heating element (4) and stabilized by digital thermostatic regulator (7) with remote sensor, attached to one of the measuring containers (5). Current system is introduced due to optimal Chlorella vulgaris Beijer cultivating temperature range of 36.5–37.5 °C.

Inner units are connected with ports (11), located in the separation walls 2 cm above the bottom. The case is filled with water at the same level as measuring containers (5).

Aerator (10) is designed to accomplish two functions:

1. air distribution into each container to provide Chlorella cultivation;
2. air agitation; air mixture is supplied by pipes with diffusive-end pipes.

Deferential breaker (8) usage assures fail-safe functioning of the facility.

Test samples of suspension from each container (5) were registered every 3 hours, using spectrophotometer SF-256. We investigated optical density of Chlorella suspension for yield gain occurrence and compared temperatures by thermometers.

The results we obtained were processed in Microsoft Excel, the average was calculated. Afterwards, readings were used to receive the optical density-time relation (figure 2).

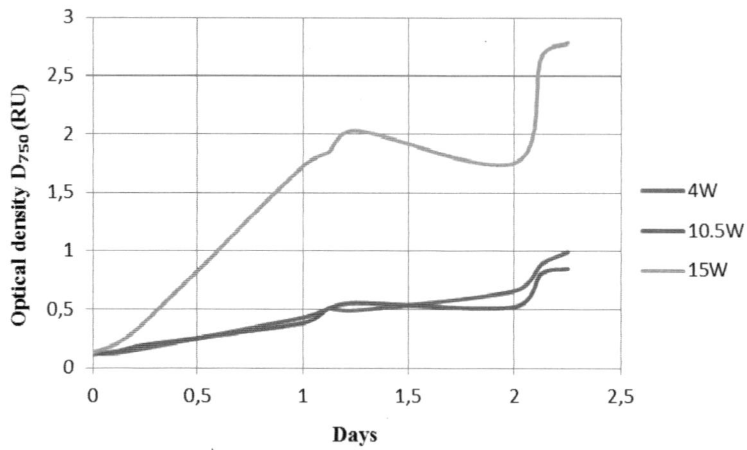


**Figure 2.** Diurnal variation of optical density at 750 nm $D_{750}$ for 3 radiation sources of different power (container 1 – 4W; container 2 – 10.5W; container 3 – 15W).

The degradation of optical density on days 1 and 2 was caused by continual light emission and stress factor absence. Furthermore, some cells died out partly and precipitated. In the day time light sources and aerator were deenergized within sampling, what lead to stress initiation and ongoing growth of Chlorella.

To reduce the measuring error while defining optical density, containers were held in flask clamps and placed as close as possible to the photoelectric detector. The detector is designed for translucent medium measurement by standard. In turbid medium light dispersion appears.

Data processing is not needed to show optical density related to wavelength (figure 3) [6].

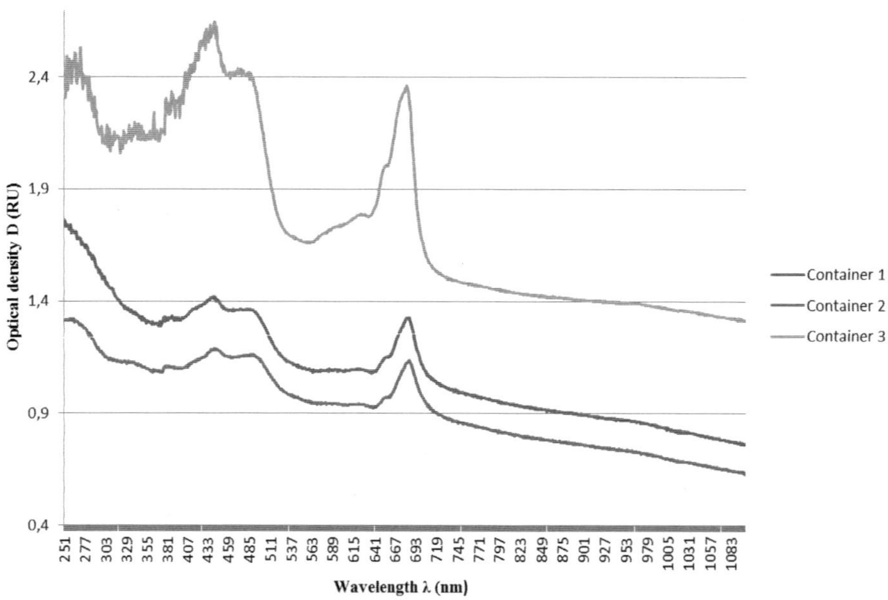


**Figure 3.** 54 hours variation of optical density for 3 radiation sources of different power (container 1 – 4W; container 2 – 10.5W; container 3 – 15W).

To determine the relation concerning cells concentration increase and time, we calculated number of cells in Gorjaev's chamber. The number of cells was calculated as,

$$n = N_{b.sq.} \cdot 2.5 \cdot 10^6 \tag{1}$$

where $n$ is the number of cells per ml, $N_{b.sq.}$ is the number of cells within Gorjaev's chamber big square.

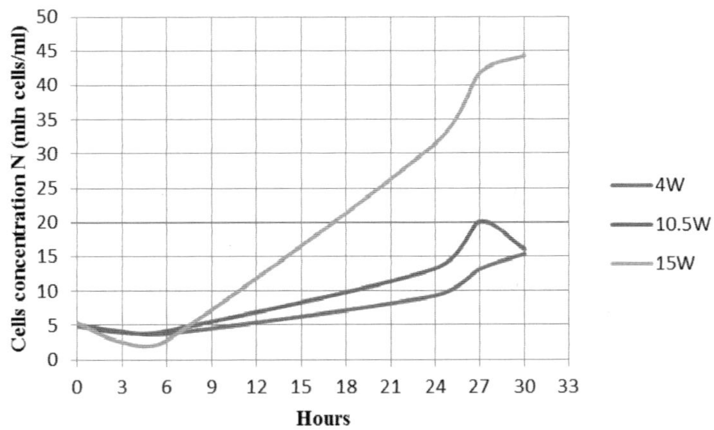


**Figure 4.** 30 hours variation of cells concentration for 3 radiation sources of different power (container 1 – 4W; container 2 – 10.5W; container 3 – 15W).

## 3. Results

Based on conducted subsidiary experiments, we proved, that with the increase of light source power level from 4W to 10.5W, without changing the spectrum of the source, optical density value multiplied by 1.5 times. The rise of light source power level from 10.5W to 50W caused 1.4 times increase of optical density. According to the results, it is not necessary for the light source power level to be at maximum to assure optimal level of algae cell growth. Maximum PAR irradiance at a small-size Chlorella suspension volume suppresses rapid growth and cell fission.

Three patterns of culture development can be identified in the figure 4: from stock culture seeding until the 6$^{th}$ hour of growth period – exponential growth phase, from the 6$^{th}$ hour until 25$^{th}$ hour of growth period – linear growth phase. Concentration decrease at the stationary phase can be observed in the figure 2. Chlorella cells fission occurs 40 times every 24 hours approximately. In accordance with this fact, cells from the second sample had not divided once again after 30 hour cultivation, when they were counted, due to lighting condition difference with the other samples.

## 4. Conclusion

Light radiation plays essential part in microalgae development, specifically light intensity and spectrum content [7]. According to the experimental data (Figures 1 - 4), highest gain of optical density and concentration was observed during 15 W LED lamp illuminance of suspension, in comparison to LED lamps with less power. This occurred as a result of photosynthesis process of microalgae: the more appropriate the conditions were, the faster cell fission speed and growth appeared.

Based on the experiment, optimized illuminance influences microalgae growth on equal terms with nutritional medium and $CO_2$ supply. The diagrams of chamber 1 and 2 show, how photosynthesis process in cells is decelerated by insufficient illuminance conditions.

## Acknowledgments

The work was supported by the Ministry of Education and Science of the Russian Federation under the State Program in the field of scientific investigations №13.3647.2017/PSH.

**References**

[1] Bogdanov N I 2007 *Chlorella suspension in live-stock animals' diet* ed L M Lepilina (Penza) 48

[2] Microalgae cultivation in photobioreactor [Web log post] Retrieved February 27, 2015, from http://biofuellab.ru/micro/fito.php

[3] Puzanskii R K, Tarakhovskaia E R, Maslov I I and Shishova M F 2011 The influence of ectogenous organic matter and illuminance on microalgae growth *Vestnik of Saint Petersburg University* **3** p 86

[4] Johannes Bialona and Thomas Rathb 2018 Growth rates and photon efficiency of Chlorella vulgaris in relation to photon absorption rates under different LED-types *Algal Research* **31** pp 204–215

[5] Alekseev M A, Ivanova S S, Korshunov K O, Malakhov A S, Trofimchuk O A and Shevchenko I G 2017 Spectra influence of three led modules on the microalgae chlorella growth in artificial conditions *Problems and prospects for the development of domestic lighting, electrical engineering and energy* p 97

[6] Gevorgiz R G and Malakhov A S 2018 *Conversion of the illumination quantity of photobioreactor into the irradiance quantity: educational methodology manual* (Sevastopol: Kovalevsky Institute of Marine Biological Research ) 60

[7] Khalili A, Najafpour G D, Amini G and Samkhaniyani F 2015 Influence of Nutrients and LED Light Intensities on Biomass Production of Microalgae Chlorella vulgaris *Biotechnology and Bioprocess Engineering* **20** pp 284–290

# The influence of the LED lighting on structural-functional parameters of lettuce plants

M S Yamburov[1], V V Volkomorov, K S Rogaev[1], E R Gorpinich[1], S B Romanova[1], M V Nevidomaya[1], E Yu Zharnakova[1], A A Burenina[1], T P Astafurova[1] and S B Turanov[2]

[1]National Research Tomsk State University, Lenin avenue 36, Tomsk, 634050, Russia
[2] National Research Tomsk Polytechnic University, 30 Lenin Avenue, Tomsk 634050, Russia

E-mail: tyrsb@yandex.ru

**Abstract.** Comparative studies were conducted on morphofunctional parameters of lettuce (*Lactuca sativa*) plants cultivated with the use of supplementary lighting from different light sources – LED lamps, high-pressure mercury arc lamp (HPM) and high-pressure sodium arc lamp (HPS). The work showed that morphofunctional parameters of lettuce were significantly higher with LED lighting, in comparison to plants cultivated with HPM illumination, and main parameters of leaves grown under LED were barely discernible from the samples grown under HPS lighting. We observed accelerated transfer into reproductive period during HPM lighting, what can lead to quality degradation of the product. Considering the approximate values of leaf character range in experiments with HPS and LED lighting and greater energy efficiency of LED, light emitting diodes are obviously more promising for supplementary illumination in protected ground conditions.

## 1. Introduction

Sun light is the ideal lighting source for plants development, though natural illumination is insufficient for heliophilous horticultural crops during winter time. It is necessary for greenhouse facilities to provide supplementary illumination to increase light flux and extend photoperiod (length of lighting). Nowadays, sodium-vapor, metal-halide lamps and luminous tubes are used for industrial lighting. Although these lighting sources are efficient for electric power- photon flux transformation, they have some disadvantages [1]. HPS lamps are widely used as supplementary lighting sources under protected ground conditions; they have long-term service and satisfactory spectrum quality. However, only 30% of electrical power, consumed by these lamps, is transformed to light emission, while 70% of it is lost for heat emanating from the power supply. The spectrum of metal-halide lamps is more suitable for plants cultivation, unlike HPS lamps. Nevertheless, just 24% of power is converted into visible radiation, what is considerably lower than HPS lamps can provide. Luminous tubes appliance in protected ground conditions is limited due to their illumination intensity, which is far behind the intensity of sodium-vapor and metal-halide lamps [2].

Modern light emitting diodes cover the whole visible range of optical spectrum, from violet to red. Light energy output of LED depends on current intensity through the crystal, what is used to adjust radiant power of the lighting source. By changing current value of different light emitting diodes, spectra with various structures and intensities can be achieved. Therefore, we can select spectrum of the illuminant, according to the exact nascent stage of the plant [3-8].

The aim of this investigation is to obtain the insight into how LED lighting affects morphofunctional and physiological parameters of lettuce leaves and compare to HPS and HPM lighting sources impacts.

## 2. Materials and methods

Lettuce (*Lactuca sativa* subsp. *secalina* Alef.) of Moskovsky Parnikivy variety was chosen as model site. Plants were cultivated in 7 L containers, filled with soil mix of acid-free turf, vermicompost and sand at the ratio of 3:1:1, maintained at the temperature range +22–25°C and water vapor saturation 70–80%. Cultivation proceeded within 12 h photoperiod, supplementary lighting was provided by different light sources. Group 1 was illuminated with LED lamp, group 2 – high-pressure mercury arc lamp (HPM; Philips 250 W), group 3 – high-pressure sodium arc lamp (HPS; Reflax 150 W).

LED system was configured to equal-energy light output for red, blue and green spectral band. LED array consisted of 1 blue, 1 green and 2 red diode contours. Blue contour included powerful high-brightness light-emitting diodes 3GR-B-5630, wavelength 445-452 nm and half-breadth of a spectral line 25 nm. Lightness – 295-340 mW during 350 mA current. Green contour contained powerful high-brightness light-emitting diodes Cree XPEGRN-L1-0000-00C01 (green), wavelength 520-535 nm. The first red contour comprised powerful high-brightness light-emitting diodes Cree XPERED-L1-0000-00501 (red), wavelength 620 nm; the second – powerful high-brightness light-emitting diodes 3GR-R, chip AlGaInP 38x38 mil (Epistar), wavelength 650-660 nm and half-breadth of a spectral line 20 nm. Lighting sources were arranged at height which ensured equal irradiance of plants throughout all 3 experiments.

The measurements of physiological and morphological parameters of plants were carried out after 30 days from the sprouts appearance. Physiological parameters determination included investigation of chlorophyll-a and chlorophyll-b sum and amount of polyphenols and valuation of nitrogen balance index. The examinations were conducted on leaves of living plants by field non-destructive testing device – flavonoid- and chlorophyllmeter Dualex (Forse-A, France). Morphological and physiological parameters of 25 differentiated leaves (1 leaf per 1 plant) were determined in each alternative experiment. We measured such morphological parameters as length, width, area, dry weight. To evaluate the area, we scanned leaves, measured each leaf area in pixels, using the program AxioVision (Carl Zeiss), and converted obtained data into $cm^2$. Leaf density per unit surface (LMA) illustrates the amount of dry matter per area unit; LMA was estimated according to the data.

Statistical processing of the data was conducted on program Statistica (StatSoft). The average values of parameter and standard deviation were calculated. Statistical significance of differences was determined by Student's t-test ($p \leq 0.05$ and $p \leq 0.01$). For leaves of plants grown under LED lighting significance of differences was evaluated with use of t-test and then contrasted with other experimental samples.

## 3. Results and discussion

Lettuce plants cultivated with LED supplementary lighting had higher values of major leaf parameters, compared to those, which were illuminated by HPM lamp (table 1). The length and width of leaves grown under LED were 1.8 times greater, area – 3 times, dry weight – 2.7 times, flavonoids sum – 1.6 times. However, plants cultivated with the use of HPM lamp had 20% higher density per unit surface (LMA) factor, 1.5 times larger chlorophylls sum and 2.3 times greater nitrogen balance index.

It is recognized, that plants with low level of LMA (thinner leaf) photosynthesize more efficiently due to less ground tissue development outgoings and light energy loss. The LMA decrease can illustrate the rise of photosynthetic active tissue ratio to the volume of cell walls in dry weight [9].

It should be noted, that 30% of lettuce plants cultivated under HPM supplementary illumination had formed flower-bearing stem and transferred into reproductive period. Possible reason for this oddity is high rate of nitrogen balance index within exact experiment. Most crops extensively consume nitrogen from the soil when approach reproductive form. At this period lettuce plants become inapplicable over their dull eating qualities.

**Table 1.** Morphological and physiological parameters of lettuce leaves cultivated under different experimental conditions.

| Factor | Lighting source | | |
|---|---|---|---|
| | LED lamp | High-pressure mercury arc lamp (HPM) | High-pressure sodium arc lamp (HPS) |
| Length, mm | 149.1±13.5 | 78.0±17.9[b] | 137.9±19.1[a] |
| Width, mm | 58.5±4.8 | 31.6±8.5[b] | 58.1±7.4 |
| Area, $cm^2$ | 45.6±5.3 | 15.6±7.2[b] | 46.2±11.1 |
| Dry weight, mg | 52.6±7.7 | 21.8±12.0[b] | 58.6±23.6 |
| LMA, $mg/cm^2$ | 1.16±0.11 | 1.38±0.25[b] | 1.26±0.29 |
| Chlorophylls sum, $mg/cm^2$ | 8.61±1.20 | 13.77±1.30[b] | 9.48±1.16[b] |
| Flavonoids sum, $mg/cm^2$ | 0.79±0.15 | 0.65±0.18[b] | 1.02±0.28[b] |
| Nitrogen balance index | 11.24±2.52 | 22.42±6.31[b] | 9.94±2.77 |

[a] Statistically significant difference to LED lighting under $p \leq 0.05$.
[b] Statistically significant difference to LED lighting under $p \leq 0.01$.

Plants grown with the use of HPS supplementary lighting had 8% less leaf length and 13% lower nitrogen balance index, but 10% greater chlorophylls sum and 29% higher flavonoids sum in contrast to LED illumination. Statistically significant differences among major parameters (width, area, dry weight, density per unit surface) were not observed.

## 4. Conclusion

Comparative investigation was carried out in lettuce (*Lactuca sativa*) plants which were cultivated under LED supplementary lighting. Morpho-physiological parameters of leaves were notably higher than one's of samples, grown under HPM lamp. Slight difference within main indexes was detected between samples with LED and HPS illumination. Considering the close values of leaf character variety of lettuce plants in experiments with HPS and LED lamps and way more energy efficient LED sources, it is obvious, light emitting diodes are more promising for supplementary illumination for protected ground conditions.


## Acknowledgement

This research was supported by "Tomsk State University competitiveness improvement program" grant (№ 8.1.29.2018).
The work was supported by the Ministry of Education and Science of the Russian Federation under the State Program in the field of scientific investigations №13.3647.2017/PP.


## References

[1] Wheeler R M 2008 A historical background of plant lighting: an introduction to the workshop *Hortscience* **43** 1942–43

[2] Wallace C and Both A J 2016 Evaluating operating characteristics of light sources for horticultural applications *Acta Horticulturae* **1134** 435–43

[3] Astafurova T, Lukash V, Goncharov A and Yurchenko V 2010 Phytotron for supplementary LED lighting of plants in greenhouses and at home *Solid-state lighting* **3** 36–38

[4] Avercheva O V, Bassarskaya E M, Zhigalova T V, Leonteva M R, Berkovich Y A, Smolyanina S O and Erokhin A N 2010 Photochemical and photophosphorylation activities of chloroplasts and leaf mesostructure in chinese cabbage plants grown under illumination with light-emitting diodes *Russian Journal of Plant Physiology* vol 57 **3** 382–91

[5] Prokofev A, Turkin A and Yakovlev A 2010 The perspectives of LED appliance i for crop farming *Solid-state lighting* **5** 60–63

[6] Singh D, Basu C, Meinhardt-Wollweber M and Roth B 2015 LEDs for energy efficient greenhouse lighting *Renewable and Sustainable Energy Reviews* **49** 139–47

[7] Berkovich Y A, Konovalova I O, Smolyanina S O, Erokhin A N, Avercheva O V, Bassarskaya E M, Kochetova G V, Zhigalova T V, Yakovleva O S and Tarakanov I G 2017 LED crop illumination inside space greenhouses *REACH - Reviews in Human Space Exploration* **6** 11–24

[8] Viršilė A, Olle M and Duchovskis P 2017 LED Lighting in Horticulture *Light Emitting Diodes for Agriculture* ed Dutta Gupta S (Singapore: Springer)

[9] Vasfilov S P 2011 The analysis of the cause of variability of the relationship between leaf dry mass and area in plants *General biology journal* **72** 436–54

IOP Conf. Series: Materials Science and Engineering **510** (2019) 012019 doi:10.1088/1757-899X/510/1/012019

# The influence of different lighting sources on growth and development of tomato plants

Mikhail S Yamburov[1], Svetlana B Romanova[1], Valentina M Smolina[1], Anastasia A Burenina[1], Tatyana P Astafurova[1], Sergey B Turanov[2] and Sofia A Romanenko[2]

[1]National Research Tomsk State University, Lenin avenue 36, Tomsk, 634050, Russia
[2]National Research Tomsk Polytechnic University, 30 Lenin Avenue, Tomsk 634050, Russia

E-mail: tyrsb@yandex.ru

**Abstract.** Comparative study was conducted on morphofunctional parameters of tomato plants (*Solanum lycopersicum* L.) under diverse supplementary lighting conditions – electrodeless lamps (EL), high-pressure mercury arc lamps (HPM) and high-pressure sodium arc lamps (HPS). The work showed, that plants, cultivated under 5,000 lux EL lighting, had higher morphometric parameters, their leaves contained more chlorophyll, and they were outgrown, in comparison to plants, growth in HPS and HPM lighting conditions. These results demonstrate the profitableness of EL light flux, which contains constant PAR spectrum, for tomato plants growth and development.

## 1. Introduction

Spectrum structure of emission has massive influence on plants development: it affects photosynthesis process, controls morphogenesis, hormonal state, ontogeny and secondary metabolism. However, results vary among different species and even breeds [1-5]. Considering that, general guidance for light spectrum appliance, appropriate for plants cultivation under photoculture conditions, is impractical [6].

According to different authors, optimum spectral structure contains blue (400–500 nm), green (500–600 nm) and red (600–700 nm) PAR radiation at the intensity ratio of 3:2:5 [7].

Nowadays, HPS and HPM lamps are widely used in major greenhouse facilities, though lighting technique market offers the latest power saving sources, which have greater luminous power and spectrum, specially selected for plants cultivation. These include electrodeless lamps (EL), where the combination of 3 physical processes – electromagnetic induction, gas discharge and phosphor glow in reaction with gas – is used to produce light emission. EL has improved technical characteristics (extended working period, greater photooptical efficiency, low heating temperature, etc.) over the other lighting sources.

The research is aimed at the examination of electrodeless lamp impact on growth and development processes of tomato plants, in comparison to HPS and HPM lamps illumination.

## 2. Equipment and methods

Tomato plant (*Solanum lycopersicum* L.) of Karamelka variety was chosen as model site. Plants were grown in Siberian Botanic Garden in moderate tropical stove at day temperature 25°C and night temperature 22°C. Seeding was performed into plastic cultivating containers of 500 $cm^3$ volume. Soil mix consisted of acid-free turf, vermicompost and sand at the ratio of 3:1:1. In each container 1 sample of tomato was grown. Containers with plants were placed on rack, equipped with different lamp types: electrodeless "Ekosvet" (EL), high-pressure sodium arc lamps (HPS) and high-pressure mercury arc lamps (HPM). Light units were arranged at height which ensured equal irradiance level. EL had constant spectrum and light flux, which contained constant PAR spectrum; HPS lamp – banded spectrum and significantly broadened D-lines of sodium with high self-reversal and maxima of 470, 483, 589 and 810 nm; HPM lamp – line spectrum and maxima of 365, 404, 435, 546 and 578 nm. As plants were growing, lamps height was raised every 2 weeks to ensure, that luminance of the upper parts of shoots maintained constant throughout the vegetation period. Diurnal supplementary lighting was 12 h. 3 versions of the experiment were conducted, and 25 plants per each variant were cultivated.

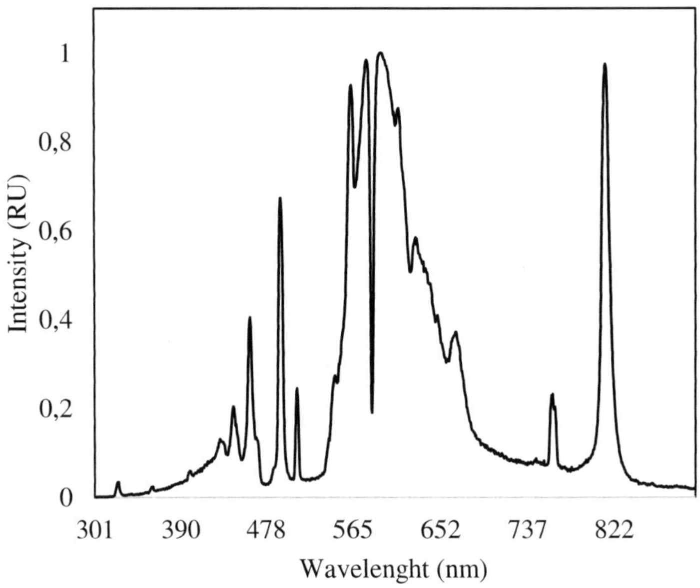


**Figure 1.** HPS lamp spectral structure.

After 70 days from sprouts rise, measurements of physiological and morphofunctional parameters of the plants were carried out. Physiological parameters determination included investigation of chlorophyll and polyphenols amounts and valuation of nitrogen balance index. The examinations were conducted on leaves of living plants by field non-destructive testing device – flavonoid- and chlorophyllmeter Dualex (Forse-A, France). Morphological and physiological parameters of 25 differentiated leaves (1 leaf per 1 plant) were determined in every alternative experiment.

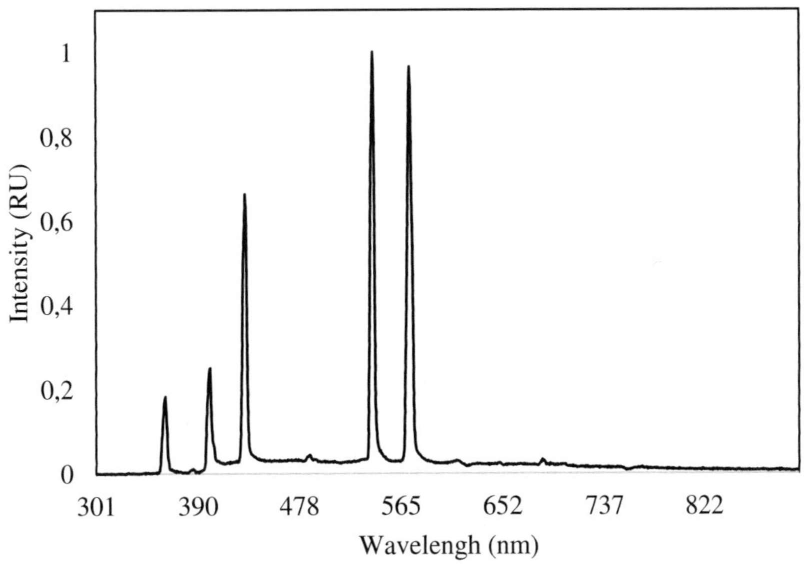


**Figure 2.** HPM lamp spectral structure.

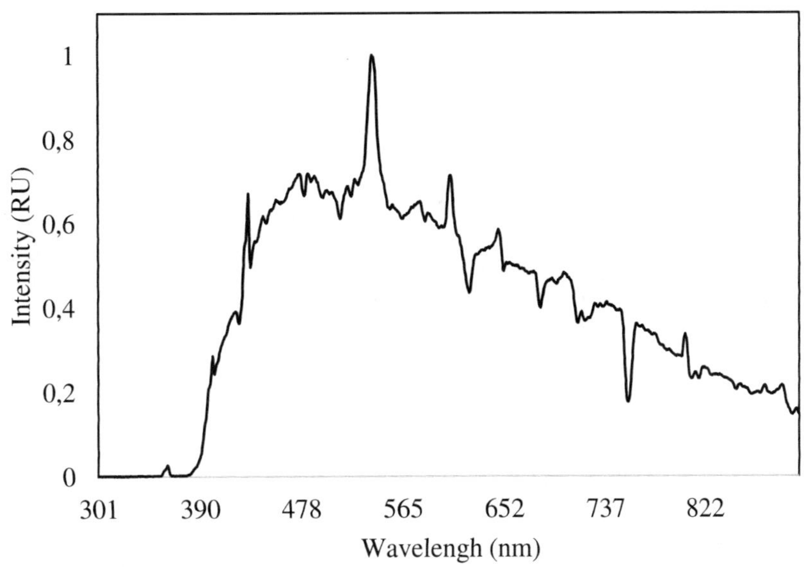


**Figure 3.** Electrodeless lamp spectral structure.

Following factors of tomato plants were defined: plant height, diameter of central section of the stem, internode length, dry weight of aerial part, leaves number on the main stem, leaf length, lamina area and leaf dry weight. To evaluate the lamina area, we scanned 2 leaves from the central section of each plant stem, measured leaf area in pixels, using the program AxioVision (Carl Zeiss), and converted obtained data into $cm^2$.

Statistical processing of the data was conducted on program Statistica (StatSoft). The average values of parameter and standard deviation were calculated. Statistical significance of differences was determined by Student's t-test ($p \leq 0.05$). For leaves of plants grown under EL lighting significance of differences was evaluated with use of t-test and then contrasted with other experimental samples.

## 3. Results and discussion

Plants, cultivated under EL supplementary lighting conditions, were more developed and outgrown on all morphometric indicators, contrasted with samples from alternative experiments (HPS and HPM). Stem diameter, height of the plant and internode length were 30% and 35% greater relatively. Moreover, dry weight of aerial part was also higher by 15 and 21 percent (see figure 4, table 1).

**Figure 4.** Plants which were cultivated in 12 h supplementary lighting conditions, provided by different lamps: 1 – electrodeless lamp (EL); 2 – high-pressure sodium arc lamps (HPS); 3 – high-pressure mercury arc lamps (HPM).

**Table 1.** Morphological and physiological parameters of tomato plants, grown in different variants of the experiment.

| Factor | Light source<br>Electrodeless lamp (EL) | High-pressure sodium arc lamps (HPS) | High-pressure mercury arc lamps (HPM) |
|---|---|---|---|
| Plant height, cm | 80.4±13.1 | 52.0±9.1[a] | 52.3±8.4[a] |
| Stem diameter, mm | 4.3±1.0 | 3.1±0.5[a] | 3.1±0.2[a] |
| Internode length, cm | 69.1±13.9 | 45.0±9.7[a] | 44.9±8.4[a] |
| Dry weight of aerial part, g | 171.2±19.8 | 145.7±22.5[a] | 134.9±23.8[a] |
| Leaves number, pcs | 11.8±1.6 | 10.3±1.5[a] | 9.9±1.3[a] |
| Leaf length, mm | 115.0±31.8 | 80.9±16.1[a] | 81.3±16.1[a] |

| | | | |
|---|---|---|---|
| Length of lobe of a leaf, mm | 56.0±15.3 | 38.5±12.7[a] | 37.2±13.2[a] |
| Lamina area, $cm^2$ | 60.0±20.3 | 25.8±8.3[a] | 22.0±8.1[a] |
| Leaf dry weight, mg | 108.8±52.6 | 54.9±18.4[a] | 59.8±28.6[a] |
| Chlorophylls sum, $mg/cm^2$ | 20.66±3.52 | 16.35±2.11[a] | 16,87±2.50[a] |
| Flavonoids sum, $mg/cm^2$ | 0.60±0.21 | 0.74±0.12[a] | 0.61±0.17 |
| Nitrogen balance index | 34 | 22[a] | 28[a] |

[a] Statistically significant difference to EL lighting under $p \leq 0.05$ (Student's t-test).

EL illuminated plants were more foliate, the number of leaves on the main stem was 13% and 16% higher than in plants illuminated by HPS and HPM relatively. Leaf length and leaf length lobe under EL lighting raised by 30 and 31–34 percent, in contrast with other variants. The increase of length, number of lobes and leaf size in EL illumination conditions refers to considerably greater value of total assimilating area of leaves. For the plants grown under HPS lamp, area and dry mass were 57% and 50% lower, under HPM lamp – 63% and 45% lower, as relevant.

The content of photosynthetic pigment chlorophyll in leaves, illuminated by EL, was considerably higher, than in alternative experiments – 21% (HPS) and 18% (HPM). The content of flavonoids between EL and HPM lamp illumination were insignificantly different for statistics, but for HPS lamp flavonoids were accumulated for 23% more intensively than in other alternatives. Nitrogen balance index of EL illuminated leaves was far greater, than for variants – 36% (HPS) and 20% (HPM). Large nitrogen balance index can be ascribed to reproductive state reaching and fruit setting under EL light conditions; it is well known, tomato plants actively consume nitrogen from soil especially during 1-7 raceme development [8].

Cultivated under EL light plants outgrew samples from alternative experiments, and by the end of examination 62% of plants achieved reproductive state (1 raceme initial blossom). Only 11% plants reached reproductive state in HPM experiment. Moreover, 38% of EL samples had off-shoots, intended to grow; off-shoots were not detected in other experiments.

We should point out, that plants cultivated with the use of HPS and HPM lamps had early aged and sallowed leaves at the stem bottom.

## 4. Conclusion

When irradiating tomato plants, electrodeless lamps, as a source of supplementary lighting, assure higher morphometric parameters of plants, content of chlorophyll in leaves and outgrowing, in comparison to HPS and HPM lamps. These results demonstrate the profitableness of EL light flux, which contains constant PAR spectrum, for tomato plants growth and development.

## Acknowledgement

The results were obtained under terms of the Russian Ministry of Education and Science state assignment, project № 37.7810.2017/8.9.

The work was supported by the Ministry of Education and Science of the Russian Federation under the State Program in the field of scientific investigations №13.3647.2017/PP

## References

[1] Lillo C and Appenroth K J 2001 Light regulation of nitrate reductase in higher plants: which photoreceptors are involved? *Plant Biology* **3** 455–65

[2] Giliberto L, Perrotta G, Pallara P, Weller J L, Fraser P D and Bramley PM 2005 Manipulation of the blue light photoreceptor cryptochrome 2 in tomato affects vegetative development, flowering time, and fruit antioxidant content *Plant Physiology* **137** 199-208

IOP Conf. Series: Materials Science and Engineering **510** (2019) 012019 doi:10.1088/1757-899X/510/1/012019


[3] Johkan M, Shoji K, Goto F, Hashida S and Yoshihara T 2010 Blue light-emitting diode light irradiation of seedlings improves seedling quality and growth after transplanting in red leaf lettuce *HortScience* **45** 1809–14

[4] Liu H, Fu Y, Hu D, Yu J and Liu H 2018 Effect of green, yellow and purple radiation on biomass, photosynthesis, morphology and soluble sugar content of leafy lettuce via spectral wavebands "knock out" *Scientia Horticulturae* **236** 10–17

[5] Chen X L, Yang Q C, Song W P, Wang L C, Guo W Z and Xue X Z 2017 Growth and nutritional properties of lettuce affected by different alternating intervals of red and blue LED irradiation *Scientia Horticulturae* **223** 44–52

[6] Berkovich Yu A, Konovalova I O, Smolyanina S O, Erokhin A N, Avercheva O V, Bassarskaya E M, Kochetova G V, Zhigalova T V, Yakovleva O S and Tarakanov I G 2017 LED crop illumination inside space greenhouses *REACH - Reviews in Human Space Exploration* **6** 11–24

[7] Bakharev I, Prokofev A, Turkin A and Yakovlev A 2010 LED lamps for greenhouses illumination *Contemporary Technologies in Automation* **2** 76-82

[8] Brovko G A 2003 *Scientific-and-practical basis of vegetable crop in static greenhouses of Russian Far East* (Vladivostok: Primpoligrafcombinat) p 180

IOP Conf. Series: Materials Science and Engineering **510** (2019) 012020 doi:10.1088/1757-899X/510/1/012020

# Mechanism of hot spots formation in pentaerythritol tetranitrate under pulsed laser irradiation

**V I Oleshko[1], V P Tsipilyov[1], V E Zarko[2], A N Yakovlev and M A Mazur[1]**

[1]National Research Tomsk Polytechnic University, 634050, Tomsk, Russia
[2]Voevodsky Institute of Chemical Kinetics and Combustion, SB RAS, Novosibirsk

E-mail: oleshko@tpu.ru

**Abstract.** The spectral, kinetic, spatial, and amplitude characteristics of emissions of pressed samples of PETN (pentaerythritole tetranitrate) by the first harmonic of a neodymium laser (1064 nm, 12 ns) are studied. It is found that at the moment of irradiation of PETN by laser pulses, hot spots are formed on the surface of the samples. The energy density of laser radiation varies in the range of 0.5–3 J/cm$^2$. The characteristics of hot spots are determined by absorbing inhomogeneities and parameters of laser radiation. Multi-pulse excitation leads to the annealing of absorbing irregularities and increased threshold of optical breakdown. The received results are interpreted on the basis of ideas of the low-threshold optical breakdown as evolving within local (defective) regions of the PETN in the electric field of the light wave.

## 1. Introduction

Laser initiation of brisant explosives opens wide perspectives for the application of the explosion energy in various technological fields and blasting technologies. These technologies have been intensely studied over the last few years. However, the initial stages of PETN explosive decomposition triggered by pulsed laser radiation (LR) are still not studied. Recently, several alternative models of laser initiation of pure (without introduced additives) PETN specimens have been suggested [1–6]. The thermal spot model is now widely used, which states the impact of power laser radiation on an explosive to be accompanied by heating of local micro-regions (defects) termed as hot spots (HS), which leads to rapid chemical transformations. However, currently, there are no reliable experimental results validating this model.

The present work is aimed at elaborating immediate experimental evidence of hot spots formation in pure PETN samples under the impact of the first harmonic of neodymium laser and determining their formation mechanism.

## 2. Methods

The studied specimens of PETN ($C_5H_8N_4O_{12}$), sugar ($C_{12}H_{22}O_{11}$) and MgO were compacted in a steel mold under pressure $P = 1.2 \cdot 10^8$ Pa into 6-mm tablets with the density of about 1.6 g/cm$^3$. The reasons to study sugar were its inertness and physicochemical properties being close to those of PETN. The mechanism of laser initiation of PETN was studied by the comparison of threshold, spacial, spectral and kinetic characteristics of the emission of the selected materials occurring under power laser impact.

The excitation source was Q-switching YAG:$Nd^{3+}$ laser at fundamental frequency ($\lambda$ = 1.06 μm) with pulse width at half maximum of 16 ns. The detailed scheme of the laser testbed for the investigation of explosive decomposition of energy-yielding materials is presented elsewhere [7]. A part of LR was deflected by glass plates to the calibrated IKT-1N calorimeter (Russia) and FEK photoreceiver (Russia) to measure energy, duration and shape of laser pulses. The density of the pulse energy *(H)* on the surface of sample was measured by QE25LP-S-MB pyroelectric energy meter (Gentec-EO, Canada) and varied using calibrated glass filters in the range of (0.1–6) J/cm$^2$. The laser beam was focused by a lens (with focal length of 0.3 m) into a 2-mm spot and incident on the plane of the open surface of sample at the angle of 75°. The experiments were performed in atmospheric air. The luminance kinetics of the excited target was recorded by H5773-04 photosensor (Hamamatsu, Japan) and DPO 3034 oscilloscope (Tektronix, USA). To record the spacial distribution of luminance along the irradiated surface of the targets, the testbed was supplemented by a microscope. The luminance induced by pulsed irradiation of the samples was captured by Sony Alpha DSLR-A500 (Sony, Japan) via MPD-10 microscope (Russia). The channels of the photomultiplier tube and photoelectric colorimeter were synchronized by the pulse of CdSe crystal two-photon luminescence excited by the first harmonic of Nd-laser. The time resolution of the measurement system was ~10 ns, while the spacial resolution was ~10 μm.

## 3. Results and Discussion

It was experimentally shown that the irradiation of pure PETN samples by the laser pulse with $H$ = (0.5–3) J/cm$^2$ is accompanied by the formation of bright spots on the background of less intense diffused luminance (Figure 1a).

(a) (b) (c)

**Figure 1.**Microphotograph of surface luminance of PETN (*a*), sugar (*b*) and MgO (*c*) under the first laser irradiation pulse. *a)* $H$ = 3 J/cm$^2$; *b)* $H$ = 2.5 J/cm$^2$; *c)* $H$ = 3.5 J/cm$^2$.

A typical oscillogram of local region (HS) luminance (upper oscillogram) in Figure 2 was obtained simultaneously with photo capturing. The lower oscillogram in Figure 2 is the pulse of Nd-laser generation. The data on spacial distribution of HSs along the irradiated PETN surface, luminance kinetic and LR oscillogram were obtained after primary LR pulse. The repeated irradiation of the surface of sample by a sequence of pulses decreased the number of HSs and their luminance intensity, i.e. HSs were annealed. The increase in $H$ from 0.5 to 3 J/cm$^2$ for irradiation of the untreated surface of the specimen increases the number of HSs, the intensity and duration of their luminance after the first excitation pulse. The HS luminance duration at pulse half maximum increases from 20 to 36 ns.

IOP Conf. Series: Materials Science and Engineering **510** (2019) 012020 doi:10.1088/1757-899X/510/1/012020

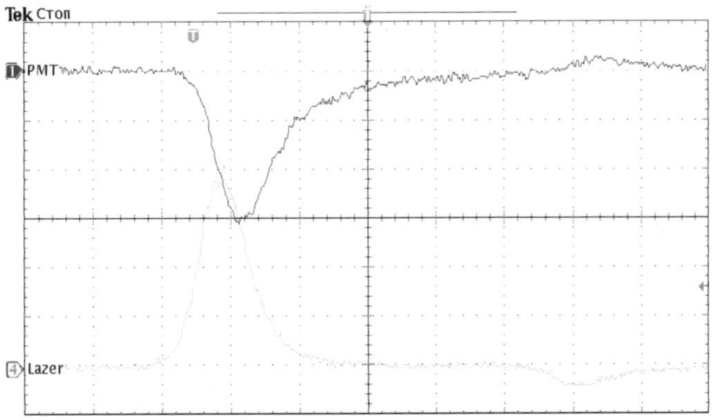


**Figure 2.** Oscillograms of hot spot luminance in PETN after the first irradiation pulse (upper) and excitation laser pulse (lower), H = 3 J/cm$^2$.

There was no damage to the PETN sample surface after the irradiation. At $H > 4$ J/cm$^2$, on the regions of the brightest HSs there are microcraters with the size of ~100 μm. Interestingly, according to early studies [1], the crater formation threshold ($H$*) in compacted PETN samples after irradiation by the first harmonic of Nd-laser depends on the PETN dispersity and varies in the range of ($4 \geq H^* \geq 9$) J/cm$^2$ with the dispersity changing from 3200 up to 12,000 cm$^2$/g. The authors of [1] have associated $H$* with the threshold of PETN optical breakdown recorded by the luminance of the laser torch and intense sound (click). However, that work lacks data on the processes taking place in PETN at $H < H^*$.

To understand the physical nature of HS luminance occurring in PETN under pulsed irradiation, additional experiments were performed using the inert samples (compacted tablets of sugar and MgO). During the studies, we have established that in the inert materials and at the energies of ($0.5 \geq H \geq 3.5$) J/cm$^2$, the regularities of formation and annealing of HSs are similar to those in PETN. All the above was confirmed experimentally and the results are presented in Figs. 1b and 1c. The results state unified character of formation and annealing of HS under pulsed irradiation of compacted samples. One can assume that HSs originate in solid bodies due to localization of LR energy on absorbing irregularities. According to [8], one of such irregularities in the compacted samples of pure PETN are carbon microparticles. To check this assumption, we have studied aged (stored for 20 years) PETN samples with 0.1 wt% of carbon. The photo images of the specimens demonstrated dark microparticles (assumably, the agglomerates of carbon particles) with the dimensions from 10 to 150 μm. We have established that at the moment of pulsed irradiation, HSs are formed in the locations of microparticles which dimensions exceed those of the particles (Figure 3a). After LR action, one can observe a light patch on the irradiated surface of sample along with the microparticles. Assumably, it is caused by the burning of soot particles having the dimensions of less than 10 μm (Figure 3b).

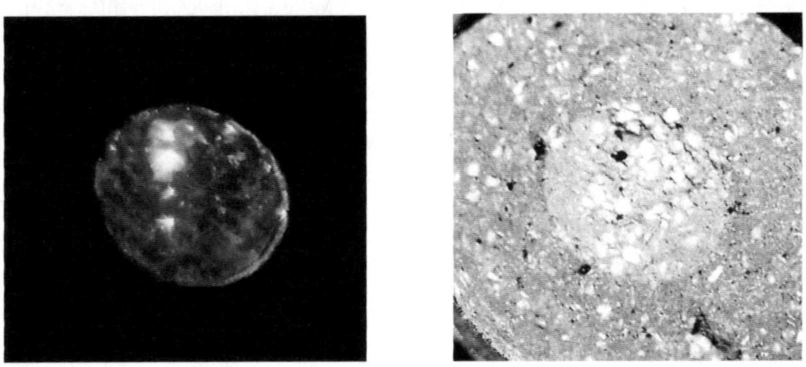

(a) (b)

**Figure 3.**Microphotographs of hot spot luminance formed in PETN subsurface region with addition of 1 wt% of carbon (*a*) and surface of sample after single irradiation (*b*).

The combination of received experimental data (nonuniform surface luminance structure with spots, delayed HS luminance maximum in relation to the laser pulse, their increased luminance duration at increased laser radiation energy density, annealing effect) and analysis of literature on the effect of absorption irregularities on the optical breakdown threshold of various materials [9–10] allows concluding that the reason of HS formation in pure PETN samples and studied inert materials is low-threshold optical breakdown evolving in local (defective) regions of solid bodies in the electric field of the light wave. The host spots are subsurface laser plasma that forms after the evolution of the optical breakdown in the vapor of absorbing irregularities. The discovery of the nature of the absorbing irregularities responsible for the initiation of the optical breakdown in the studied samples requires additional investigation. We assume that one of the origins of absorbing irregularities are impurities introduced into the specimens after the compaction in the metal mold. It is confirmed by the authors of [11] who have established the absence of HSs for the energy densities of $0.16 < H < 2.5$ J/cm$^2$ in sugar samples prepared in a glass cell. Separate microzones of white luminance, in this case, were detected only for $H \leq 2.5$ J/cm$^2$.

## 4. Conclusion

The studies have demonstrated that the excitation of the compacted samples of studied materials by the first harmonic of Nd-laser causes formation of bright hot spots (local micro- and macrozones of white luminance) forming after the development of low-threshold optical breakdown on absorbing irregularities. The role of absorbing irregularities is played by microparticles introduced into the subsurface region of the samples during their compaction in a metal mold. Multi-pulse excitation of the samples leads to the evaporation of absorbing impurities (annealing of defects) and increased threshold of optical breakdown.

Thus, we can conclude that the HS formation mechanism in the samples of pure PETN under LR pulse is electrical, similarly to the irradiation of explosives and inert materials by high-current electron beam [12–13]. Under laser action, the electrical breakdown develops in the subsurface region of the sample on absorbing irregularities in the electric field of the light wave, while under electron irradiation it happens in the zone of electron beam breakage, in the electric field of volumetric negative charge injected into the sample. Finally, the received results in addition to the scientific value bear practical interest, because the significance of the nature and parameters of the absorbing irregularities present in energy-yielding materials and possibility of their annealing under sequential irradiation enable the adjustment of the explosive material sensitivity to laser radiation.

### Acknowledgement

The work was funded within the grant of the Programme for Promoting the Competitiveness of Tomsk Polytechnic University, project No. VIU_IFVT_73/2017, and partially under the support of agreement No. 15592 as of October 1, 2018.

### References

[1] Tarzhanov V I 1998 *Collection of publications* (Snezhinsk: Press of All-Russian Scientific Research Institute of Technical Physics) p 126 (in Russian)
[2] Bourne N K 2001 *R. Soc. Lond. A.*. **457** 1401-26
[3] Kuklja M M, Aduev B P and Aluker E D 2000 *J. Appl. Phys.* **89 7** 4156-66
[4] Aluker E D, Krechetov A G, Loboyko B G et at. 2008 *Chem. Physics*. **27 5** 67-70. (in Russian)
[5] Aluker E D, Belokurov G M, Krechetov A G et al. 2010 *Technical Physics Letters* **36 6** 80-85 (in Russian)
[6] Aduev B P, Nurmukhametov D R, Leeskov I Yu et al. 2014 *Combustion and Explosion Physics.* **50 1** 124-129 (in Russian)

[7] Tsipilev V P 2003 *Bulletin of the Tomsk Polytechnic University* **306** **4** 99-103 (in Russian)
[8] Chernai A V 1996 *Combustion and Explosion Physics* **32** **1** 11-19 (in Russian)
[9] Vorobyov V S 1993 *Physics-Uspekhi.* **163** **12** 51-83 (in Russian)
[10] Assovskiy I G, edit. By A M Lipanov 2005 *Combustion physics and internal ballistics* (M.: Nauka) p 357
[11] Oleshko V I, Tsipilev V P, Yakovlev A N, Murastov G V, Alekseev N A 2018 *Optics and Spectroscopy* **124** **6** 804-807 (in Russian)
[12] Oleshko V I, Damamme G, Malys D, Lisitsyn V M 2009 *Technical Physics Letters* **35** **20** 55-61 (in Russian)
[13] Oleshko V I, Lisitsyn V M, Skripin A S, Tsipilev V P 2012 *Technical Physics Letters* **38** **9** 37-43 (in Russian)

# Organization of information exchange between digital companies of Industry 4.0

**D A Zakoldaev[1], A V Shukalov[1], I O Zharinov[1] and O O Zharinov[2]**

[1] Faculty of Information Security and Computer Technologies, Saint Petersburg National Research University of Information Technologies, Mechanics and Optics, 49, Kronverksky Av., Saint Petersburg, 197101, Russia
[2] Department of Problem-Oriented Computing Complexes, Saint Petersburg State University of Aerospace Instrumentation, 67, Bolshaya Morskaia str., Saint Petersburg, 190000, Russia

E-mail: mpbva@mail.ru

**Abstract**. The paper studies an urgent task of organizing the information exchange between design, production and exploitation companies of Industry 4.0. The information exchange is based on different types and principles of deployment. The access to cloud services for the designer, production operator and exploitation operator is based on wireless technology of industrial Internet of Things. The work suggests the interaction scheme and describes the purpose of the information exchange components between Industry 4.0 companies.

## 1. Introduction

Information support for design, production and exploitation of equipment parts is based on the implementation of advanced digital technologies at each stage of the article life cycle. The advanced digital technologies of Industry 4.0 include additive manufacturing, cloud technologies, Industrial Internet of Things (IIoT) and other [1-3].

To integrate digital technologies, one needs a unified information space between design, production and exploitation companies. Design and production companies of Industry 4.0 are digital and smart factories that are equipped with cyber-physical systems (CPS) [4, 5]. CPS enable automatic article creation implementing their digital models and digital twins of technological equipment [6, 7].

The exploitation companies of Industry 4.0 involved into item life cycle are united into clusters forming a virtual factory. A virtual factory is to account and control item movement during the exploitation including the information support of logistics (item delivery to the exploitation destination, item installation at the facility), warehouse storage, item sending to the manufacturer for repair and so on [8, 9].

Thus, the task of developing a scheme of information exchange between the companies of the future—digital factories, smart factories and virtual factories—is high on the agenda to guarantee the non-interrupted life cycle of an equipment part.

## 2. Physical-world components for interaction of Industry 4.0 companies

All infrastructure components that are necessary for Industry 4.0 companies to interact are of two types: physical and virtual. Physical components are hardware equipment of design and production and exploitation processes of the items [10 ,11]. The physical world components of an Industry 4.0 company are given in Figure 1.

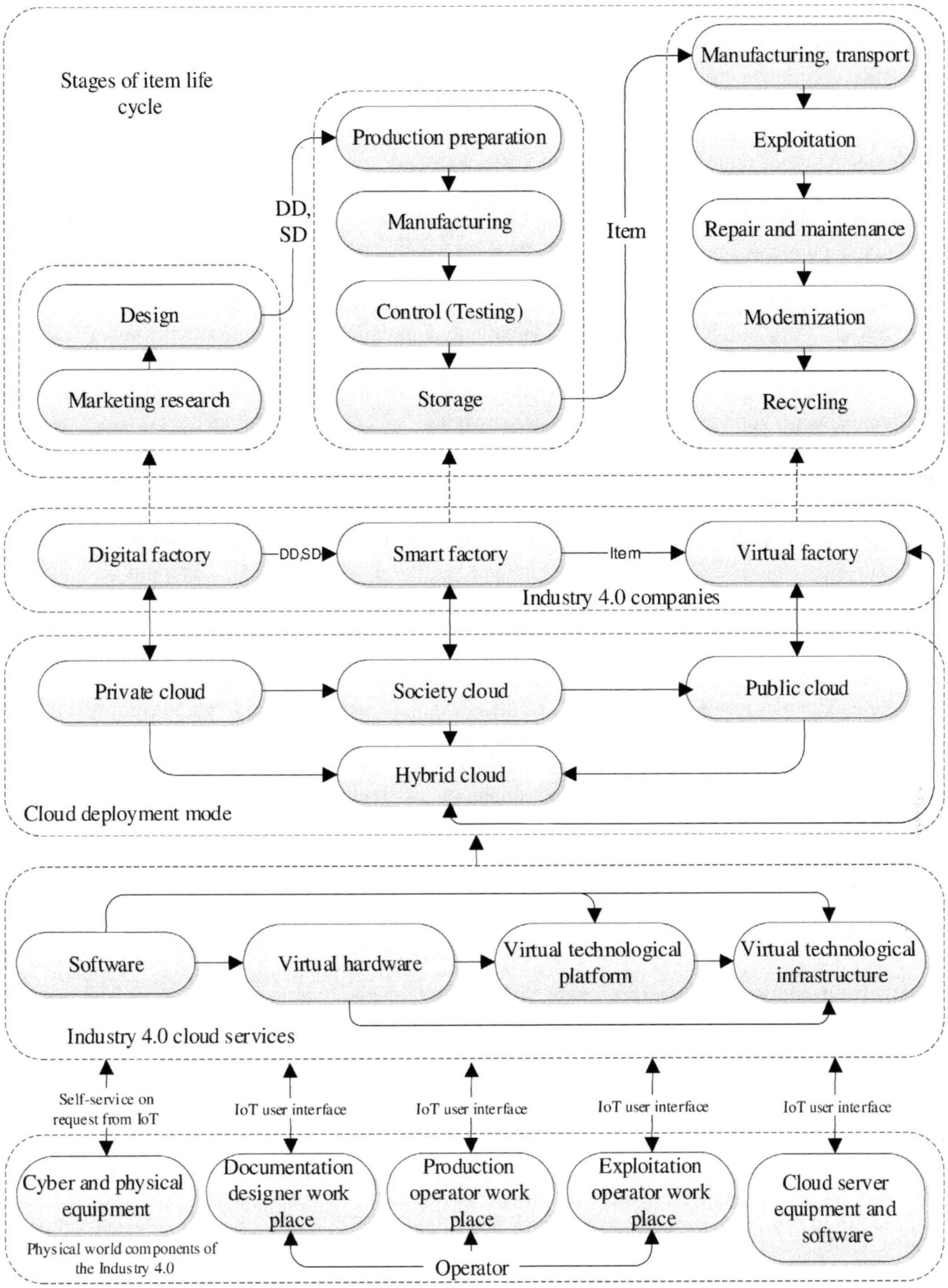


**Figure 1.** Interaction scheme of information exchange components between Industry 4.0 companies.

The physical-world components of an Industry 4.0 company are:

- cyber-physical systems, engaged in the production process at smart factories;
- automated work places of the developers of design documentation (DD), software documentation (SD) and technological documentation of items designed at digital factories;
- automated work places of production operator for quality assessment and to control the stages of equipment part manufacturing at smart factories;
- automated work places of exploitation operator at a virtual factory to monitor and control the movement of equipment part during exploitation;
- server equipment (computers, switches, routers, data storages and other) and software which on the physical level creates cloud services with remote access.

The operators of the automated work places interact with cloud services via user interface software by IoT protocols. The means of human-to-machine communication are as follows: personal computer, tablet, phone or other equipment with a screen, browser and access to internet. An Industry 4.0 company includes many automated work places synchronized with each other to perform design and production operations in line with the part development of or its manufacturing algorithm.

Cyber-physical systems [4, 5] enable automatic information exchange with cloud services components with *self-service on request* access provided by IoT components. Each component of a cyber-physical system has personal software-supported internet access port in a unified channel of cloud information exchange for the mechanical assembly line.

## 3. Virtual-world components for interacting Industry 4.0 companies

Software services create virtual components with remote user access through an automated work place. The cloud services necessary for information exchange between the Industry 4.0 companies are given in Figure 1.

Cloud services of Industry 4.0 are [1-3]:

- program-type service which provides a digital factory designer with the final piece of software to design an item and automatically create template-based technical documentation;
- computer-type service which gives a digital factory designer and smart (virtual) factory operator the final piece of hardware for computation, technical calculations, modelling, rendering, user software installation and other. Virtual calculation machine is a software server available to user as a resource;
- platform-type service which gives a digital factory designer and smart (virtual) factory operator a means to develop special user applications destined for particular design and production operations. Consequently, the ready-to-use components of user applications can be regarded by a user as a program-type cloud service;
- infrastructure-type service which gives a digital factory designer and smart (virtual) factory operator a means to manage digital engineering and financial data about company activity, digital models of manufactured equipment components, company digital twins and its technological equipment (cyber-physical systems).

Obviously, the infrastructure-type service is the most efficient service and is created after the integration of lower-tier services: program, computer, and platform. The integration of services (pooling of cloud services) can be done by several means (deployment models) in a single cloud that is available to the personnel of the factories of the future.

Cloud deployment models are defined by the tasks of a particular Industry 4.0 company and can be:

- a private cloud which corresponds to the activities of a single company which performs a particular set of life stages of an equipment part. Such company is an Industry 4.0 digital factory which carries out some design tasks to create technical documentation for an equipment part;
- a community cloud which corresponds to the activities of a group of *supplier-consumer* companies and unified document workflow within the production cooperation. Such group of companies is called smart factory, which manufactures items of equipment and their components (materials, parts);
- public cloud which corresponds to the activities of a group of companies (virtual factory) which personnel is to monitor the condition of an equipment item during its life cycle (from the stage of marketing research until the stage of recycling);

- hybrid cloud that in terms of the user access type unites hardware and software solutions and services used in a private cloud, community cloud or public cloud.

Thus, the cloud service deployment model defines the means and right of access for an Industry 4.0 company personnel to different cloud services.

The data storages of a private cloud contain the following components as ready-to-use Computer-Aided Design (CAD) solutions:

- archive of design solutions generated automatically by the developers as a part of a digital factory operation. The archive of digital factory project solutions includes newly developed and already existing solutions with high unification and standardization. The design solutions are digital models of equipment items;
- archive of tasks for the development of an item of equipment and its parts (private development tasks);
- archive of company standards which regulates the processes of design and electronic document flow in the company;
- archive of reference and technical regulation documentation which is necessary to develop technical documentation for equipment items according to the industrial and state standards;
- archive of technical documentation for standard items and materials (resistors, capacitors, metals, rubber, paint and other) used as raw materials and components to design new equipment;
- mathematical models of environment (externalities) necessary for certain stages of virtual tests on digital models of equipment items performed at a design company. Virtual tests are done to evaluate the compliance of an item under development with the technical task (technical solutions suggested by designers are validated using mathematical models, while the design results are rendered on the computer screen of an automated work place).

The cloud storage of contains the smart factory data as library services:

- archive of company standards which regulate the production operations and electronic documentation flow in the company;
- workflow charts and technical documentation for equipment item provided to a company for its mass production;
- algorithms of technological processes and operations which consequently are used to compose a workflow chart equipment item production;
- digital twins of technological equipment (cyber-physical systems) installed at a production company and enabling automatic production of equipment items;
- models of components and materials which are used in the company to create an equipment item;
- network models and charts of equipment item manufacturing at a company, including production data of both primary manufacturer and suppliers of components working within the production cooperation.

The cloud data storage of a virtual factory contains the library services of:

- archive of the standards regulating the electronic document flow in the company;
- data on the operation of the item and its operation documentation;
- list of contractors that contribute to the item life cycle (development, production, transportation, maintenance, repair, and other).

In Figure 1, the dashed arrows show the connection between Industry 4.0 companies and the item life cycle stages implemented via cloud services.

## 4. Conclusion

The development of modern item design industry inherently includes the improvement of technological basis of design and production companies and organizational support of companies that use the item. The united information space is made available to the personnel of the factories of the future through a set of services deployed in a cloud.

The cloud service contains a set of data that corresponds to mathematical, software, information, linguistic, methodical and technical support which interaction is regulated via organizational support.

The organizational support application by the personnel of the factories of the future enables the information exchange between Industry 4.0 companies at each stage of equipment item design.

**References**

[1] Aljoumah E, Mousawi F, Ahmad I, Al-Shammri M and Al-Jady Z 2015 *International Journal of Grid Distribution Computing* **8** 7-32
[2] Zanoon N, Al-Haj A, Khwaldeh S M 2017 *International Journal of Applied Engineering Research* **12** 6970-82
[3] Charan N R G, Rao S T, Srinivas P V S 2011 *International Journal of Advanced Computer Science and Applications* **2** 119-125
[4] Chen J, Yang J, Zhou H, Xiang H, Zhu Zh, Li Y, Lee Ch-H and Xu G 2015 *Engineering* **1** 247-260
[5] Zhang Ch, Jiang P, Cheng K, Xu X W and Ma Y 2016 *Procedia CIRP* **56** 360-365
[6] Liu Ch, Jiang P 2016 *Procedia CIRP* **56** 372-377
[7] Leitao P, Colombo A W, Karnouskos S 2016 *Computers in Industry* **81** 11-25
[8] Petnga L, Austin M 2016 *Advanced Engineering Informatics* **30** 77-94
[9] Delgoshaei P, Austin M A, Pertzborn A J 2014 *International Journal on Advances in Systems and Measurements* **7** 223-238
[10] Wang Y, Liu D, Sun Ch A 2017 *Energies* **10** 267
[11] Cai Y, Starly B, Cohen P, Lee Y-Sh 2017 *Procedia Manufacturing* **10** 1031-42

IOP Conf. Series: Materials Science and Engineering **510** (2019) 012022 doi:10.1088/1757-899X/510/1/012022

# Kinematical analysis of the nutation speed reducer

**M A Jasem and P Y Krauinsh**

Department of materials science, School of Advanced Manufacturing Technologies, Tomsk Polytechnic University, Timakova street, 12,Tomsk, 634028, Russia

E-mail: eng.mjasem85@gmail.com, peterkrau@tpu.ru

**Abstract**. This paper discusses the development of a Nutating Speed Reducer (NSR) which is characterized by high reduction ratio, high tooth contact ratio, very high torque to weight/volume ratio, quiet and smooth operation under load and very high efficiency. All of these advantages are due to the presence of conjugate face-gear pairs, which incorporate each other, which called nutating/rotating gear mechanism. Details of the NSR, its kinematics, gear tooth load capacity, and mesh efficiency are explained. The NSR component speeds and speed reduction ratios of the NSR are calculated. Effect of the varying nutation angles on the geometry of the NSR is discussed and compared.

## 1. A brief history of the nutating gear

Robert Davison discovered the concept of nutation almost by accident [1]. While he was throwing a coin on the table, he noticed that as a speed of spinning coin begins to decrease, a peculiar wobble takes place. This wobble is nutation. This concept has been adapted to design high gear speed reductions, which is known as "Nutating Gear Device".

Nutating gears are high shaft angle bevel gears where one bevel gear is wobbling instead of rotating against a fixed gear. The rotation of the nutating gear depends on a difference in tooth numbers between the nutating gear and fixed gear instead of a difference in circumferences between them.

## 2. Introduction

Nutation speed reducer is a totally innovative gear transmission developed in recent years [2, 3], which bases its functioning on the mathematical concept of nutation motion, that is the main feature that characterizes nutation gear from conventional gear. It has unique features which include high reduction ratio in one stage, very high profile contact ratio and an overall higher power density transmission. In addition, the contact is theoretically pure rolling in the gear, the heat generated should be lower and the efficiency higher.

## 3. Structure of NSR

The Nutation Speed Reducer is illustrated in Figure 1. The component names illustrated in Figure 1 are (1) an input shaft; (2) the stator gear; (3) the nutator gear; (4) the rotor gear and (5) the output shaft.

IOP Conf. Series: Materials Science and Engineering **510** (2019) 012022 doi:10.1088/1757-899X/510/1/012022

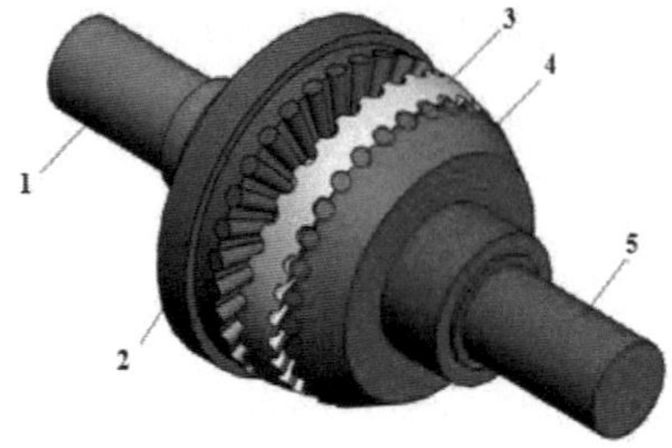


**Figure 1.** The Nutation Speed Reducer

The input shaft is driven by an engine and drives the nutator gear, which is fixed to the input shaft at a specified nutation angle (Typical range of the nutation angle $\beta$ is 2-5° from vertical direction) [4]. The nutator has gear faces on either side of the body that are in mesh with both the stator and the rotor gears. The motion and torque transfer from nutator to rotor gear and output shaft.

The numbers of teeth on the stator gear 2 and rotor gear 4 is $Z_1$, $Z_4$ respectively. The numbers of teeth on the nutator gear 3 meshing with stator gear is $Z_2$. The numbers of teeth on the nutator gear 3 meshing with rotor gear is $Z_3$. The gears of NSR simultaneously share the tooth loads with high contact ratios as a result of its unique assembly [5, 6].

## 4. Kinematics of NSR

As illustrated in Figure 2, input shaft, stator gear and rotor gear rotate about the Z-axis of the global fixed-coordinate frame Sf. The nutator gear converts the purely rotational motion of the input shaft into both rotational motion about the Z axis and nutational motion (or precession motion), which is a rotation about the $z_N$ axis. This second rotation is caused by the meshing engagement with the stator gear. The $z_N$ axis is fixed to the input shaft; so, it rotates about the Z axis.

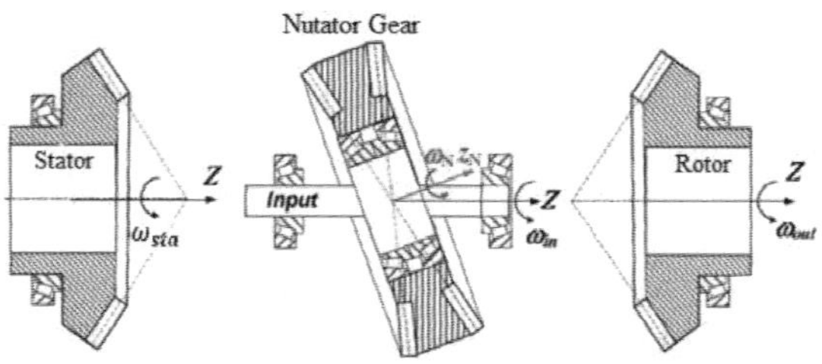


**Figure 2.** Motions of the NSR gears

The nutation speed reducer velocities are defined by three major characteristics, the nutation angle of the nutator gear, the number of teeth in the two respective meshes between the nutator and the stator, and the nutator and the rotor, and the rotational velocity applied to the stator gear (when driving the stator gear via and external power input as in Elmoznino's work [7]). In this work the stator gear is fixed to body of reducer, so the gear tooth numbers and nutation angle will be the major factors in the velocity terms [8].

Coordinate relations of the $S_f$ and $S_N$ are illustrated in Figure 3. The $S_f$ and $S_N$ coordinate frames are coincident at the common origins $O_f$ and $O_N$. The directions of the frames are set according to the right hand rule.

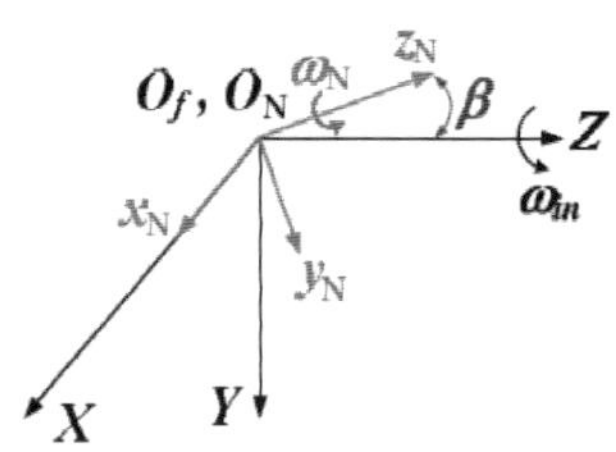

IOP Conf. Series: Materials Science and Engineering **510** (2019) 012022 doi:10.1088/1757-899X/510/1/012022

**Figure 3.** Coordinate relations of the nutator gear and global frame

Assuming an input speed for the input carrier being $\omega_{in}$, we can formulate the rotational speed of each body. The input shaft rotational velocity can be defined as:

$$\vec{\omega}_{in} = \omega_{in} \cdot \vec{K} = \begin{bmatrix} 0 \\ 0 \\ \omega_{in} \end{bmatrix} \qquad (1)$$

Where the symbol $\vec{K}$ presents unit vector in the Z direction of the $S_f$ coordinate. The rotor angular velocity is:

$$\vec{\omega}_{out} = \omega_{out} \cdot \vec{K} = \begin{bmatrix} 0 \\ 0 \\ \omega_{out} \end{bmatrix} \qquad (2)$$

The motion of the nutator gear is a combination of the input speed ($\omega_{in}$) and the speed component from the meshing engagement between the stator gear and the nutator gear ($\omega_N$). It is:

$$\vec{\omega}_{Nut} = \omega_{in} \cdot \vec{K} + \omega_N \cdot \vec{k} \qquad (3)$$

Where the symbol $\vec{k}$ is the unit vector in the z direction of the $S_N$ coordinates. The $\omega_N$ gives as:

$$\vec{\omega}_N = (\omega_{Sta} - \omega_{in})\frac{Z_1}{Z_2}\vec{k} \qquad (4)$$

A coordinate transformation is necessary to present nutator gear angular velocity in the SN coordinate. As illustrated in Figure 4 to form $S_N$ frame the $S_f$ coordinate is rotated about the X axis. The coordinate transformation matrix defining a rotation of $\beta$ degrees about the X axis is:

$$R_x = \begin{bmatrix} 1 & 0 & 0 \\ 0 & \cos\beta & \sin\beta \\ 0 & -\sin\beta & \cos\beta \end{bmatrix} \qquad (5)$$

and the transformation of the coordinate is established by:

$$\begin{bmatrix} I \\ J \\ K \end{bmatrix} = \begin{bmatrix} 1 & 0 & 0 \\ 0 & \cos\beta & \sin\beta \\ 0 & -\sin\beta & \cos\beta \end{bmatrix} \cdot \begin{bmatrix} i \\ j \\ k \end{bmatrix} \qquad (6)$$

The input speed component of the nutator gear angular velocity is transformed as:

$$\vec{\omega}_{Nut} = \omega_{in} \cdot \left(R_x \cdot \vec{k}\right) + \omega_N \cdot \vec{k} \qquad (7)$$

with expanding:

$$\vec{\omega}_{Nut} = \begin{bmatrix} 0 \\ -\omega_{in}\sin\beta \\ \omega_{in}\cos\beta \end{bmatrix} + \begin{bmatrix} 0 \\ 0 \\ \omega_N \end{bmatrix} \qquad (8)$$

and the sum of the angular velocity component vectors results in :

$$\vec{\omega}_{Nut} = \begin{bmatrix} 0 \\ -\omega_{in}\sin\beta \\ \omega_{in}\cos\beta + \omega_N \end{bmatrix} \qquad (9)$$

with a more compact form:

$$\vec{\omega}_{PMC} = -\omega_{in}\sin\beta\,\vec{j} + (\omega_{in}\cos\beta + \omega_N)\vec{k} \quad (10)$$

The nutator gear angular velocity with the number of teeth relation is then becomes:

$$\vec{\omega}_{Nut} = \begin{bmatrix} 0 \\ -\omega_{in}\sin\beta \\ \omega_{in}\cos\beta + -\omega_{in}\frac{Z_1}{Z_2} \end{bmatrix} \quad (11)$$

## 5. Speed Reduction Ratio

A simple schematic of the NSR is illustrated in Figure 4. Here, the symbols $Z_1$, $Z_2$, $Z_3$ and $Z_4$ present the numbers of teeth on the stator, nutator and rotor gears, respectively.

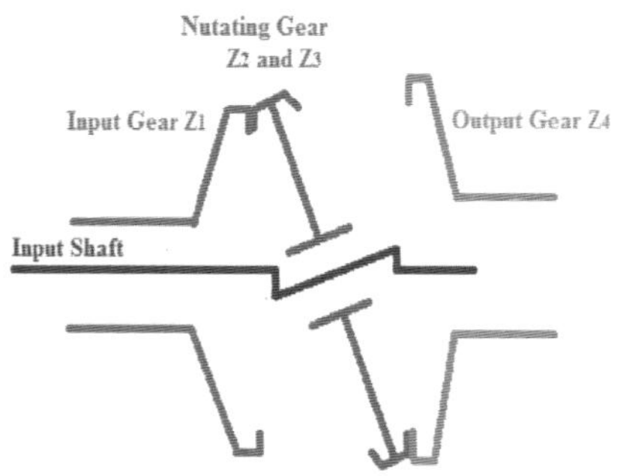


**Figure 4.** Simple schematic of the NSR

The reduction ratio ($i$) is presented as [9]:

$$i = \frac{\omega_{in}}{\omega_{out}} = \frac{1}{1 - \frac{Z_1 \cdot Z_3}{Z_2 \cdot Z_4}} \quad (12)$$

## 6. Nutator Gear Geometry

The nutator gear geometry is highly dependent upon its nutation angle, which defines the internal bevel gear pitch cones size and shape [9]. To calculate the size and shape of pitch cones, the number of gear teeth, the diametral pitch, and nutation angle has to be given. All of the pitch cone vertexes within the NSR align at the point, that the pitch cone vertexes of the two gear meshes within the PMC are coincident of one another and back to back. This is the point about which the nutator both rotates and nutates during its meshing cycle. As illustrated in Figure 5 can be seen this point with the pitch cones being shown and coming together.

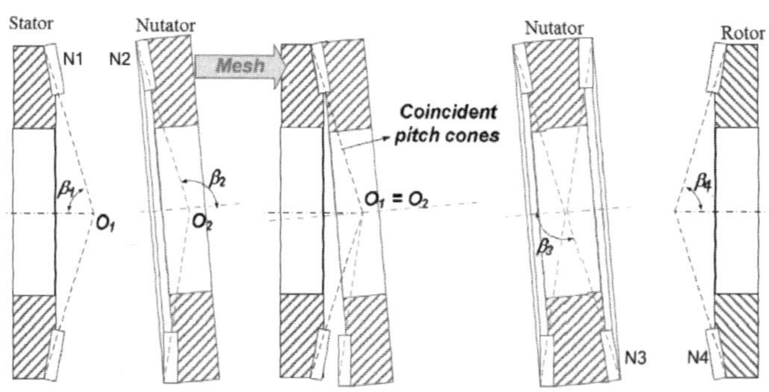


**Figure 5.** Meshing between the stator and nutator

The pitch cone angles of these three bodies are mathematically defined by five parameters, the teeth numbers of each gear pair ($Z_1$,$Z_2$,$Z_3$,$Z_4$) and the nutation angle β. To find the pitch cone angles of the four gear pitch cones using trigonometric relations:

$$\beta_1 = \tan^{-1}\left(\frac{\sin(\pi-\beta)}{\frac{Z_2}{Z_1}+\cos(\pi-\beta)}\right) \quad (13) \quad , \quad \beta_2 = \pi - \beta - \beta_1 \quad (14)$$

$$\beta_3 = \tan^{-1}\left(\frac{\sin(\pi-\beta)}{\frac{Z_4}{Z_3}+\cos(\pi-\beta)}\right) \quad (15) \quad , \quad \beta_4 = \pi - \beta - \beta_3 \quad (16)$$

## 7. Effect of Altering Nutation Angle on NSR Geometry

The variation in nutation angle impacts all pitch cone angles while not changing the reduction ratio of the NSR, therefore it is a good candidate to examine. Using the geometric parameters given in Table 1 and altering the nutation angle, we will examine how the nutator gear size is affected.

**Table 1.** Basic parameters of Nutation Speed Reducer

| Parameters | Z1 | Z2 | Z3 | Z4 | i | Module | Face Width |
|---|---|---|---|---|---|---|---|
| Value | 42 | 44 | 81 | 80 | 30 | 4.23 mm | 20 mm |

The nutator gear geometry has two main dimensions, the radial dimension from the nutator center, and the axial dimension from the nutator axis. Since the nutator has two sides, so four dimensions with varying nutation angle from 1 degree to 15 degrees were created, the resulting calculated values are illustrated in Figure 6 below.

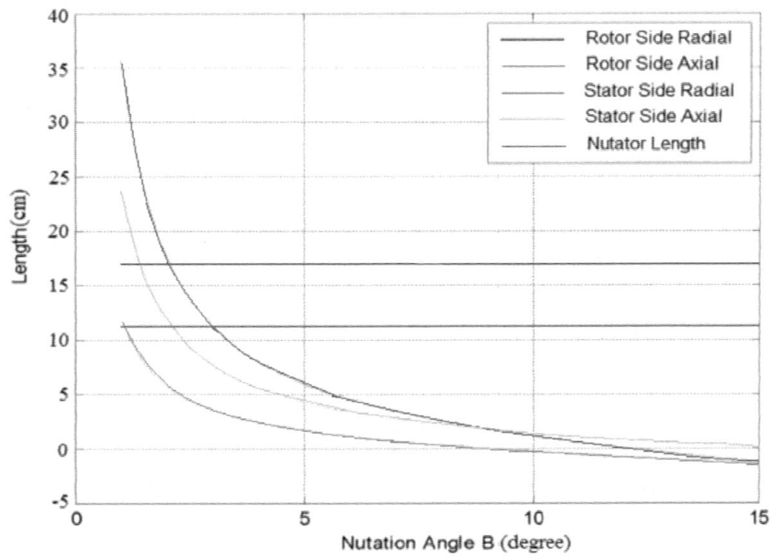


**Figure 6.** Nutator gear size against varying nutation angles

We can observe that there are no change in radial lengths, and this is due to their value being tied to the diametral pitch and number of teeth on each gear, which is held constant in this case. While increasing nutation angle causes a decrease in axial dimensions as stator and rotor side axial and nutator length. This exponential decrease in axial length immediately translate to an exponential decrease in mass of the nutator body. But the axial decrease is limited, due to it decreases the conformity of the gear meshes as well as the number of teeth simultaneously in contact, and limiting the area available for the nutator bearings. At approximately twelve degrees nutation the nutator length turns negative. Because of an inversion of the pitch cones where the NSR transfers from the speed reducer regime to a humpage drive regime [10]

## 8. Kinematic Simulation

The virtual assembling for the 3D models of NSR is carried out in a virtual environment as shown in figure 7. To verify the transmission theory and the interference detection of the nutation speed reducer, the kinematic simulation were verified. It should be noted that, the motion constraint is 3D contact pair[5] between the stator and nutator ,as well as the rotor and nutator. It can be proven from the simulation that there are no interferences between the movable parts.

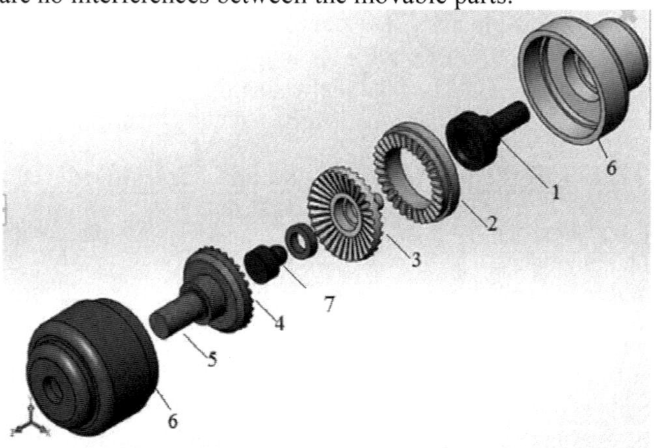


**Figure 7.** Assembly exploded views of NSR

## 9. Conclusion

In this work, a novel and advanced transmission which called Nutation Speed Reducer has been studied. The kinematical analysis of the nutation speed reducer was performed, which showed that the motion of the nutator gear is combination of two motions produced by the input shaft and stator gear mesh. These coupled motions produce the kinematics to the NSR system.

The general formulae for calculating the reduction ratio and main design parameters of the novel gear drive were developed. The geometry of the nutator gear was examined along with the impact of nutation angle on overall size and shape. The kinematic simulation was carried out in a virtual simulation environment to verify the transmission theory and the interference detection of the nutation speed reducer.

## References

[1] Kedrowski K and Slimak P 1993 *Wobbling Gear Drivetrain for Cordless Screwdriver* Honors Theses (Michigan: Western Michigan University)
[2] Lemanski A and Monahan T J 2004 Non-Traction Pericyclic CVTs *SAE International* **40** 5
[3] Nelson C A and Cipra R 2005 J *Journal of Mechanical Design* **127** 269-277.
[4] Saribay Z and Bill R 2012 Design Analysis of Pericyclic Mechanical Transmission System *Mechanism and Machine Theory* **61** 102-122
[5] Lemanski A J 2004 Variable Speed Power Transmission System *U.S. Patent* No. 6,764,423
[6] Molyneux W G 1997 The internal bevel gear and its application *Journal of Aerospace Engineering* **211** 39-61
[7] Elmoznino M 2008 The Nature of Power Flow Through the Pericyclic Continuously Variable Transmission M.Sc. Thesis (Connecticut: University of Connecticut)
[8] Binney D A 2001 *Generation of Face-gear Geometry for Kinematic Study and Loaded Analysis* Master Thesis ( Ohio: Ohio State University)

[9] Kedrowski D K and Slimak S P 1993 Wobbling gear drivetrain for cordless screwdriver *American Society of Mechanical Engineers Winter Meeting*
[10] Humpage T and Herbert I J 1901 Toothed gearing *U.S. Patent* No. 676,234

IOP Conf. Series: Materials Science and Engineering **510** (2019) 012023 doi:10.1088/1757-899X/510/1/012023

# Polycondensation processes occurring during 3D aerosol printing with new ceramic materials based on nanostructured highly dispersed fillers of polyaluminosilicates used in nano- and microelectronics and medicine

**A A Ivanov[1], V A Polyushko[2] and A Yu Khomyakov[1]**
[1]Tomsk State University of Control Systems and Radio Electronics, 40 Lenin Avenue, 634050, Tomsk, Russia
[2] Tomsk State University, 36 Lenin Avenue, 634050, Tomsk, Russia

E-mail: alexchemtsu@rambler.ru

**Abstract.** A new low temperature ceramic material for 3D printing based on polyaluminosilicate has been synthesized. Some of the synthesis stages are controlled by the parameters set for 3D aerosol printing. The processes of 3D aerosol printing of polyaluminosilicates filled with a highly dispersed filler have been studied. An optimal method for aerosol generation has been chosen that does not affect the supramolecular structure of a polyaluminosilicate binder. Ceramic dielectric topologies with high thermal conductivity and layer thickness in the nanometer range have been obtained by means of 3D aerosol printing of the synthesized material for the first time.

## 1. Introduction

3D printing is currently a rapidly developing modern technology to form various coatings and topologies in microelectronics products and devices that have related prospects for industrial application. The development of fundamental approaches to produce consumable products (the so-called "composite materials") with the necessary functional and structural properties and careful physicochemical, morphological, rheological, and metrological analysis of the obtained stable suspensions and their components is one of the main problems related to the microprinting technology development. This problem is to a large extent limited to the synthesis and chemical modification of various functionalized nanoparticles with the set chemical and phase composition and morphology and also to the study of such particles behavior in the initial solution and in the process of solvent removing from a limited volume of a microdroplet. Nowadays, many high-tech companies all over the world switch to the technology of prototyping and commercial production of modern microelectronic, sensory, and photovoltaic devices using computer design and microprinting. It is not possible to obtain a specific dielectric layer or topology with the set thermophysical, physical, and chemical properties at the certain area of a device or product by means of traditional methods with the use of commonly applied materials.

Thus, metallic and nonmetallic materials commonly applied in various industries have to a great extent reached their functional and structural limits. However, the modern microelectronics industry requires new materials reliably operating in a complex combination of force and temperature fields and being exposed to corrosive environment, radiation, extreme vacuum, and high pressure. One can solve this problem using modern nanostructured composite materials and processing techniques in combination with conventional materials.

The modern and available method to produce functional and structural materials with the necessary properties is currently a sol-gel technology that allows controlling the chemical nature of the material in the synthesis phase, thereby setting the required properties for the finished composites. The sol-gel technology is the primary method to obtain multipurpose nanodispersed materials. The actual tendency of sol-gel synthesis is to modify the siliceous compounds by both organic and inorganic substances. The sol-gel method peculiarity is the hydrolytic polycondensation reaction of alkaline silicates that occurs in the presence of inorganic substances (salt, acid) and low and high molecular organic modifiers. They give the specified physicochemical and technically essential properties to the formed materials. As a rule, the inorganic substances and modifiers are template agents contributing to the formation of silicate and hybrid nanocomposites with a specific structure [1-3].

The primary task is to apply the obtained material on different in their chemical nature bases and study the processes that occur when applying the obtained material. This task is to be solved along with the basic task to produce new ceramic composite materials with specified functional and structural properties used as consumable products for 3D printing of various coatings and topologies in microelectronic devices and products. Such process task as layers and topologies application generates current scientific challenges of surface physical chemistry, for example, the problem of material adhesive power and strength. Simultaneously, applying material on the product or device surface generates fundamental research of physicochemical processes.

Currently there are 2D and 3D printing of composite materials that relate to printing technology. The printing technology development is based on the achievements in the field of nanotechnology and nanomaterials. Consequently, the production of functional elements with the use of a printer relates to nanoelectronics. Research and development in this field of nanotechnology is a crucial task because it is dynamically developing now [1, 4-7].

The composite materials are promising because the structural elements with the phase composition peculiarities and unique properties can be formed in them. Such materials are used along with the 3D printing method to obtain a ceramic multilayered coating with high adhesive power, electric strength, and high thermal conductivity.

The study objective is to:

- research the polycondensation processes occurring during 3D aerosol printing with new ceramic materials based on polyaluminosilicates nanostructured with highly dispersed fillers;
- obtain multilayered ceramic layers and topologies of low temperature synthesis by 3D printing;
- study the test sample functional and structural properties;
- estimate their probable use as a dielectric ceramic coating with high adhesive power, electric strength, and high thermal conductivity in nano- and microelectronics, light engineering, and devices for special purposes.

## 2. Experiment

### *2.1. 3D printing. Ceramic layer printing on an aluminium base with a screw feeder.*

A screw feeder is used in those cases when dose quality, accuracy, and frequency are important. One widely uses them to apply composite materials and glue. One can get the dose with a diameter of 200 microns and less with high frequency using a screw feeder. Ceramic layers on an aluminium base have been obtained by screw dosing with the help of the Spectrum II S2-920 dosing unit (one-track version) produced by Asymtek.

A screw or Archimedean screw is the main element that influences the quality of dose formation. The material is fed from a syringe into the channel with the screw by compressed air. Then the material is forced through a needle onto the substrate surface by the normalized screw rotation. The screw rotation angle and internal needle diameter determine the dose volume. Both heads are equipped with motors with encoders and the feedback system that allows setting the rotation angle with an accuracy of 0.0095 degrees per one encoder value.

The cartridge of the screw feeder is removable. The screw rotates inside the cartridge. This allows fast replacing and cleaning of the screw and reducing the time for technological maintenance. The screw is made of tungsten carbide that has high strength, durability, and service life. It does not rust and is not exposed to oxidation.

The screw can have different screw pitches that allows using the composites of the third type and composite materials having smaller particle size. The smaller is the screw pitch, the smaller is the material dose that can be obtained. At the same time, the gap between the screw and the channel walls, through which the material is fed, is very important. It should be of such a size that the highly dispersed filler of the composite material is not pressed by bypassing the screw turns that may cause the leakage of the composite material applied on an aluminium base and the screw rotation blocking.

In addition, the cartridges differ in the type of a needle used. It is possible to use plastic, metal, and high-precision needles both with and without a stopper.

The advantages of a screw feeder:

- dosing of viscous materials such as highly filled composite materials;
- high stability and accuracy of doses;
- easy replacement and maintenance of a screw cartridge;
- the wide range of applied materials;
- a built-in encoder and the possibility of accurate control of the screw rotation angle.

The disadvantages:

- mechanical contact with a circuit board while dosing (for the needle with a stopper);
- the impossibility to apply layers and topologies with a layer thickness of less than 1 micron;
- the impossibility to apply the composite material with a wetting angle of 180°.

Figure 1 shows the profilogram of the topologies obtained as a result of the experiment.

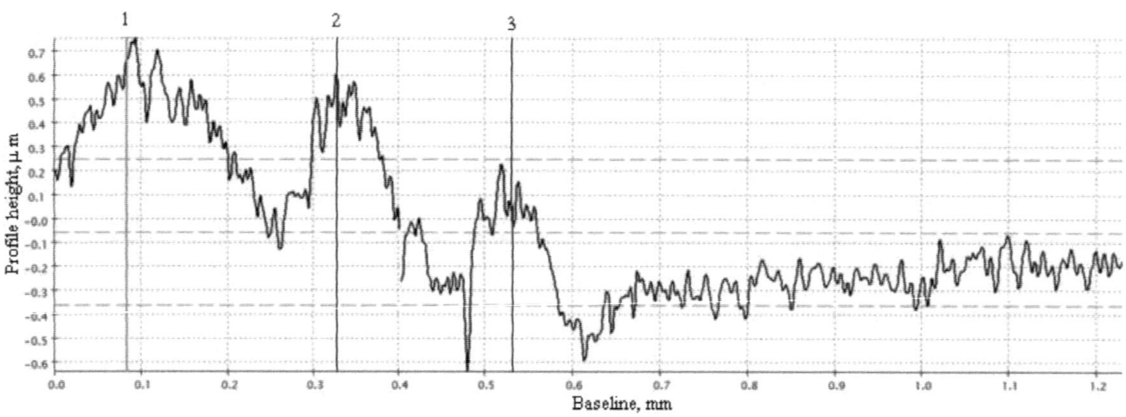


**Figure 1.** Profile profilogram of three separate topologies obtained by method of screw dosing of polyaluminosilicate filled with highly dispersed filler on aluminum base.

The profilogram in Figure 1 shows that the wetting angle of all three lines applied on a smooth surface does not meet the requirements of printed electronics that exerts a great influence on this method applicability to obtain layers and topologies from synthesized polyaluminosilicate filled with highly dispersed fillers.

*2.2. Ceramic layer printing on an aluminium base by means of 3D aerosol printing.*

IOP Conf. Series: Materials Science and Engineering **510** (2019) 012023 doi:10.1088/1757-899X/510/1/012023

To obtain aerosol from dispersed liquids is one of the most crucial problems solved in some technological processes and definitely in 3D aerosol printing.

The dispersing can be carried out by various methods, the most widespread of which are hydraulic, mechanical, pneumatic, and electrostatic.

The hydraulic method is based on fluid passing through an aperture. Pressure differential, fluid viscosity, air density, and turbulence are the main factors that affect dispersing. In this method the agglomerate strength depends on the strength of individual particle contacts, coordination number, porosity, and the size of initial particles. The synthesized composite material based on the polyaluminosilicate filled with a highly dispersed filler forms an agglomerate in the form of a ball. The ball consists of many particles streamlined by a stream being exposed to the crushing effect of the pressure differential in a face part. The lateral efforts that arise in this process separate the particles of highly dispersed filler and the polyaluminosilicate macromolecules as a whole. Since the particles are connected by autohesion forces, the separation voltage is proportional to the pressure differential. When smaller particles streamline the agglomerate, the agglomerate body may be destroyed by erosion, or vice versa, the agglomerate may capture small particles. Therefore, the aerosol from partially destroyed polyaluminosilicate is obtained at the outlet. Even partially destroyed polyaluminosilicate is not capable of forming the supramolecular structure that binds the particles of the highly dispersed filler. Consequently, the opportunity to control and set the properties of the obtained layers and topologies based on the synthesized composite material fails.

The mechanical method uses crushing by a rotating disk. The initial pressure, degree of carrier medium compressibility, concentration, material dispersion, and polyaluminosilicate binder characteristics determine the disk movement parameters. The disk structural characteristics that affect the diffusion in the obtained aerosol stream and the carrier medium expansion are crucial. The effect of the carrier medium compressibility in case of sufficient pressure differentials results in a gas-dynamic mode of two-phase medium motion and the appearance of critical conditions for the outflow that differ from the critical conditions for the compressible homogeneous medium. Thus, the force action between the particles of the highly dispersed filler and polyaluminosilicate binder in their relative motion mainly occurs in the form of surface resistance forces and to some extent in the form of the Magnus forces during particle rotation. The decrease in the particle size of the highly dispersed filler corresponds to the increase in its resistance coefficient. At the same time, the particle surface area increases per the particle mass unit that leads to the increase in the surface forces of interaction with the carrier medium, i.e. with polyaluminosilicate. Simultaneously, higher movement uniformity of the composite material in the form of aerosol is achieved. The increase in the particle size leads to relative weakening of the surface forces that involve particles in pulsating motion. At the same time, the inertial resistance to this motion increases due to the mass increase. The feedback effect of impurities on the large-scale turbulence intensity that appears while crushing by a rotating disk reduces the turbulence intensity. It is due to the fact that the energy balance of the pulsating motion and required efficiency of turbulent stresses are determined according to an increase in the average mass density of the medium involved in the pulsating motion. The mass increases when the impurity concentration of the solid highly dispersed filler particles rises. During the research this method has shown that the interdomain space between polyaluminosilicate macromolecules changes. The entire internal structure of the polyaluminosilicate filled with the highly dispersed filler also changes when being applied ("printed") with an aerosol. This fact is due to a mechanical influence that negatively affects the properties of the finished layer or topology.

The pneumatic method fragments liquid and droplets by the air flow using the droplet instability in motion. The criterion used to estimate the droplet instability determined by the ratio of the inertial liquid forces to the surface tension is called the Weber number.

If the inertial forces exceed the surface tension forces by six times, the droplets fragment into smaller ones.

The representations about the two-phase medium motion can be obtained in some limiting cases. A coarse material in limited concentrations less likely demonstrates the collisions of particles in the pneumatic method. The influence of turbulent pulsations on their motion is excluded due to the large mass of particles. As a result, it is possible to study in detail the motion of separate particles in the carrier stream by the action of forces from the stream with the simultaneous action of gravitational forces and the shock interaction of the filler particles with the container walls that occurs during motion. In this case, it is possible to obtain the data on the translational and angular velocity of the filler particles and conditions for their collision with the container wall. Moreover, it is also possible to estimate the energy costs taking place due to the loss by the wall collision or equivalent energy costs at the steady stream to restore the velocity of the filler particles after the shock that ultimately determines the additional hydraulic resistance by generating aerosol.

The distribution of particles in the container will be uniform for the highly dispersed filler particles at relatively high concentrations. In this case, the assumptions about the ideal properties of the carrier medium, i.e. polyaluminosilicate, devoid of viscosity are used to study the features of the dynamic stream modes of the compressible two-phase medium. The interaction of the carrier medium with impurity solid particles is supposed to be due to their resistance during their relative motion in the stream according to the experimentally obtained data. The additional friction resistance due to the interaction of impurities with the container wall is taken into account on the basis of the experimental data. Besides, the integral conditions for the continuity preservation, momentum, and energy in a closed form with respect to the unknown stream parameters can be determined for a compressible heterogeneous stream on this basis.

The ultrasound crushing of liquid occurs due to the energy of mechanical vibrations of the ultrasonic frequency of high intensity. R. L. Peskin determined a relationship between the film thickness δ, the amplitude α, and the frequency ώ of the exciting force. The formed drop radius for large δ is defined as follows:

$$2a = \left(\frac{4\pi^3 \cdot \sigma}{\rho_l \cdot \omega^2}\right)^{1/3}.$$

The electrostatic spraying of particles is possible when a charged liquid particle moves along the force lines of the electrostatic field under the impact of the Coulomb forces. The entire charge is distributed over the surface in a single drop of liquid. The interaction forces between the charged particle and field by the particle motion may lead to the drop crushing into smaller parts if the electrostatic field has the sufficiently high strength.

All these methods have a number of significant disadvantages. The pneumatic method is characterized by the low coefficient of transfer material being sprayed onto a product. The mechanical and hydraulic methods are characterized by high heterogeneity of the obtained aerosol. The electrostatic spraying of highly viscous liquids requires the use of an additional spraying agent. These disadvantages result in a decrease in the efficiency of technological processes, in which these spraying methods are used.

According to the research results, the most preferable method for us is the pneumatic method that allows working with highly viscous materials without destroying the supramolecular structure of a polymer binder.

We have used the synthesized polyaluminosilicate filled with a highly dispersed filler and the Aerosol Jet 15EX 3D printer of Neotech AMT company to obtain ceramic layers and topologies on aluminum bases by 3D aerosol printing with the help of the pneumatic method.

The technology to obtain and apply the aerosol is considered in the article. The applied material (polyaluminosilicate filled with a highly dispersed filler/ceramic ink) is injected into an aerosol generator. A pneumatic or ultrasonic generator can be used according to the ink type. A small amount of ink is enough to refill the ultrasonic generator, so it can be used to apply expensive materials.

Nevertheless, the range of ink viscosity and maximum diameter of solid particles are substantially limited. The pneumatic generator allows applying the materials with a wide range of viscosity and sequence larger diameter of solid particles, but a larger amount of ink is required to refill it. A working gas (nitrogen or air) is injected into the pneumatic generator through the narrow Venturi hole under pressure to produce the aerosol. The pressure increase causes the ink rise along the channel, and the aerosol is produced by the gas-ink contact. The exit pressure decreases, the gas velocity increases significantly. Consequently, fluid is sucked into the reduced pressure area through narrow channels from a chamber reservoir. When the liquid meets the air stream by the gas jet action, it is broken into small particles. These particles vary in size from 15 to 500 microns, it is the so-called "primary" aerosol. The particles then collide with a "flap" (plate, ball, etc.) that results in the "secondary" aerosol formation in the form of ultrafine particles ranging in size from 0.5 to 10 µm (about 0.5% of the initial aerosol). The secondary aerosol is then inhaled, and a large proportion of the primary aerosol particles (about 99 %) is deposited on the inner walls of the chamber and is drawn back in the aerosol formation. The aerosol jet is focused at a distance of up to 5-15 mm from a nozzle that allows applying the ink on a three-dimensional base. It can be technically achieved by moving the printing head along three axes (x, y, z) and by inclinating the base along two axes. The ultraviolet, infrared, or thermal drying is carried out after the ceramic ink is applied. The drying type depends on the number of layers and the base, which they are applied on.

The profilometric measurements of the topologies based on the polyaluminosilicate filled with a highly dispersed filler (ceramic ink) applied on a substrate have been carried out to estimate the uniformity of their application (Figure 2).

The profile measurements of the topologies printed on the substrate have been carried out with the Bruker DektakXT's device. The device measures the roughness profile and parameters by a centre line system according to the range of values. The device feels the unevenness of the measured surface with a probe (a diamond needle) moving the inductive sensor along the measured surface. Then the mechanical oscillations of the probe convert into a digital signal.

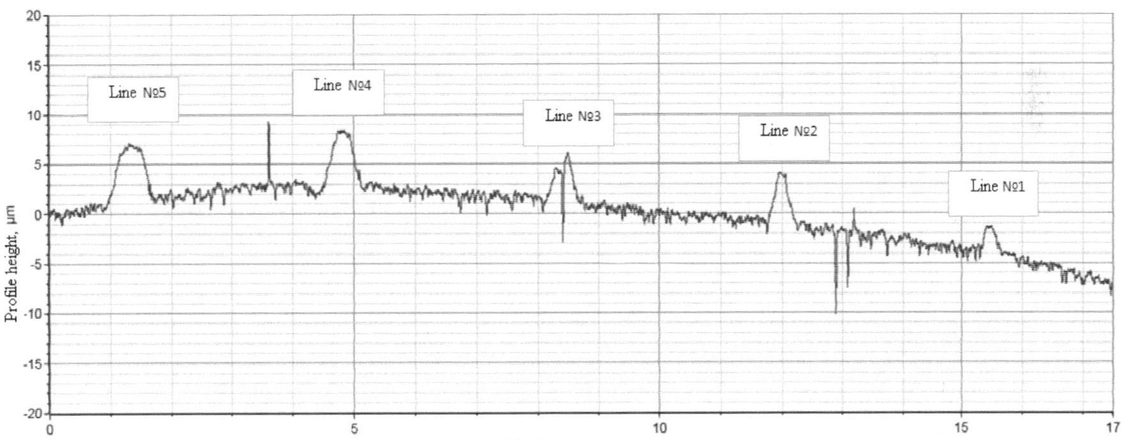


**Figure 2.** Profilogram photo of measured topologies of polyaluminosilicate filled with highly dispersed filler applied on substrate after thermal drying.

Figure 2 shows that the wetting angle changes not more than 5% (taking into account the measuring device error) depending on the number of printed layers (topology thickness) that is permissible in microelectronics manufacturing.

One of the most important parameters of the applied layers and topologies is their adhesive strength to the substrate surface, which they are applied on. A research method to study their adhesive strength has been chosen experimentally. The thin layers and small width of topologies printed with ceramic inks have been taken into account. The acoustic emission method has turned out to be the most suitable.

## 3. Conclusion

A new method of low-temperature ceramic material synthesis for 3D aerosol printing based on polyaluminosilicate has been developed. Some synthesis stages are controlled by the parameters set in 3D aerosol printing. The synthesized polyaluminosilicate matrix is able to locate up to 80% of highly dispersed filler in its inter-crown spaces and interdomain voids, thereby setting the necessary properties of the finished layers and topologies. The processes of 3D aerosol printing of polyaluminosilicates filled with highly dispersed filler have been studied. An optimal method of aerosol generation has been chosen that does not affect the supramolecular structure of the polyaluminosilicate binder. A study and comparative analysis of the 3D aerosol printing method and method of screw dosing have been carried out. The prospects of using the 3D aerosol printing method have been proved. The ceramic dielectric topologies with high thermal conductivity and thickness of layers in the nanometer range have been obtained by 3D aerosol printing of the synthesized material on the basis of the obtained results for the first time. Therefore, the use of the synthesized material based on polyaluminosilicate filled with a highly dispersed filler together with the 3D aerosol printing method allows obtaining the layers and topologies in the nano- and micrometer range that expands the nano- and microelectronics industry potential as a whole.

**Acknowledgments**

The reported study was funded by RFBR according to the research project № 18-29-11018\18.

## References

[1] Ivanov A A, Tuev V I and Vilisov A A 2016 *Key Engineering Materials* **712** 188–192
[2] Ivanov A A, Botvin V V and Filimoshkin A G 2014 *Russian Journal of Applied Chemistry* **87(2)** 141–150
[3] Ivanov A A, Botvin V V and Filimoshkin A G 2014 *Russian Journal of Applied Chemistry* **87(2)** 135–140
[4] Oth C and Kuhn T 2015 *Kunststoffe International* 76–80
[5] Khan S, Doh Y H, Khan A, Rahman A, Choi K H and Kim D S 2011 *Current Applied Physics* **11(1 SUPPL)** 271–279
[6] Awais M N, Kim H C, Doh Y H and Choi K H 2013 *Thin Solid Films* **536** 308–312
[7] Duraisamy N, Muhammad N M, Kim H and Choi K 2012 *Thin Solid Films* **520(15)** 5070-5074

IOP Conf. Series: Materials Science and Engineering **510** (2019) 012024 doi:10.1088/1757-899X/510/1/012024

# Computer simulation of the energy spectra of PKA in materials irradiated by protons in the framework of the Cascade-Probabilistic method

**T A Shmygaleva[1], A I Kupchishin [1,2], A A Kupchishin [2] and C A Shafii[1]**
[1]Kazakh National University named after al-Farabi, al-Farabi av 71, Almaty, Kazakhstan
[2]Kazakh National Pedagogical University named after Abai, Dostyk 13, Almaty, Kazakhstan

E-mail: ankupchishin@mail.ru

**Abstract.** The work is performed within the framework of a cascade-probabilistic method, the relevance of which is beyond doubt, since the models obtained make it possible to trace the entire process in dynamics and can later be used in industry to obtain materials with predefined properties. Algorithms for calculating the spectra of primary knocked-on atoms for protons have been developed. The features of the selection of approximate expressions and approximation coefficients for proton irradiation are considered. Calculations have been made for the concentrations of vacancy clusters in copper irradiated by protons.

## 1. Introduction

Modeling the radiation damage of solids irradiated with charged particles is one of the important problems of solid-state radiation physics [1, 2]. Studies in the field of radiation influence on the physical properties of materials are of great interest both for studying the structure of a solid body and for the practical use of substances with new properties obtained as a result of irradiation [3, 4].
In the interaction of charged particles with matter, primary, secondary, knocked-off atoms, etc., are generated. As a result, the structure of a solid is disturbed and radiation defects are formed. [5 – 11]. During proton irradiation, defects are generated in the form of cascade regions of the vacancy type. The main characteristic of the damaged material is the energy spectrum of the primary knocked-on atoms at different depths of the irradiated material, the knowledge of which makes it possible to calculate the number of defects formed, and also to model their distributions in depth and in dimensions, i.e. it is possible to trace quite effectively the entire process in dynamics at any depth of the irradiated material.

In this paper, models of the PKA spectra and cluster concentrations were obtained, and their depth calculations were made in materials irradiated with protons. A comparison with the previously obtained results was made. The work was performed within the framework of a cascade-probabilistic method.

## 2. Method of calculation

To calculate the spectra of primary knocked-on atoms in materials irradiated with protons, it is first necessary to find analytic expressions for cascade-probability functions (CPFs). To calculate the CPF it is necessary to obtain an approximate expression that describes the cross section of the interaction that enters into the recurrence relation, from which the expressions for the CPF are obtained. For protons, the dependence of the approximation function on energy, which in turn depends on the depth of penetration, is represented in the following form [8]:

$$\sigma(h) = \sigma_0\left(\frac{1}{a(E'_0 - kh)} + 1\right). \tag{1}$$

where $\sigma_0$, $a$, $E_0'$, $k$ are the approximation coefficients, h is the penetration depth of the particle.

The cascade-probability function with allowance for energy losses for protons is represented in the following form [8]:

$$\psi_n(h', h, E_0) = \frac{1}{n!\lambda_0^n}\left(\frac{E_0'-kh'}{E_0'-kh}\right)^{-l}\exp\left(-\frac{h-h'}{\lambda_0}\right) * \left[h - h' + \frac{\ln\left(\frac{E_0'-kh'}{E_0'-kh}\right)}{ak}\right]^n, \tag{2}$$

where $l = 1/(\lambda_0 ak)$, $\lambda_0 = 1/\sigma_0$, h', $h$ are the generation and registration depths of the proton, respectively, $n$ is the number of interactions, $E_0$ is the initial energy of the proton.

The spectrum of primary knocked-on atoms is determined by the formula:

$$W(E_0, E_2, h) = \frac{E_d E_{2\max}}{E_2^2(E_{2\max} - E_d)}\sum_{n=0}^{n_1}\int_0^h \exp(-\ln n! - n\ln\lambda_0 - \frac{1}{\lambda_0 ak}\ln\left(\frac{E_0'}{E_0'-kh'}\right) - \frac{h'}{\lambda_0} +$$

$$n\ln\left(h' + \frac{\ln\left(\frac{E_0'}{E_0'-kh'}\right)}{ak}\right) - \frac{h-h'}{\lambda_2} - \ln\lambda_1(h') - \ln\lambda_2)dh'', \tag{3}$$

where $E_d$ is the threshold energy of displacement, $E_{2max}$ is the maximum possible energy acquired by the atom, $\lambda_1(h')$ and $\lambda_2$ are the ranges of displacement of the proton and atom-atom displacements, $n_1$ is the maximum number of collisions during proton irradiation. The concentration of radiation clusters during proton irradiation is calculated by the formula:

$$C_k(E_0, h) = \int_{E_c}^{E_{2\max}} W(E_0, E_2, h)dE_2\,, \tag{4}$$

$E_2$ is the energy of the primary knocked-on atom, and $E_c$ is the threshold energy for cluster formation.

## 3. Results and discussion

The integrated cross-section of the interaction was calculated by Rutherford's formula [12], the penetration depths were obtained from the tables of ranges and braking powers for light ions [13]. For the calculated cross-section of the interaction, the approximate expression was chosen by formula (1) depending on the depth of observation, and the approximation parameters included in the expression for CPF and PKA spectra were found. The results of the approximation coefficients are given in Table 1 and in Figure 1.

**Table 1.** Approximation values and theoretical correlation ratios for protons in Cu

| $E_0$, (MeV) | $\sigma_0$ | a | $E_0$' | K | η |
|---|---|---|---|---|---|
| 5 | 3341 | 0.17 | 4.11 | 500.4 | 0.9997 |
| 10 | 1810 | 0.0445 | 18.45 | 704.4 | 0.9998 |
| 15 | 1973 | 2.8 | 0.74 | 15.01 | 0.999 |
| 20 | 1087 | 0.55 | 2.08 | 24.64 | 0.999 |
| 25 | 821 | 0.0954 | 10.77 | 85.47 | 0.9997 |
| 30 | 779 | 0.24 | 5.51 | 32.39 | 0.999 |
| 40 | 1097 | 0.14 | 34.34 | 126.6 | 0.99 |
| 50 | 947.7 | 0.12 | 49.13 | 122.6 | 0.99 0.99 |

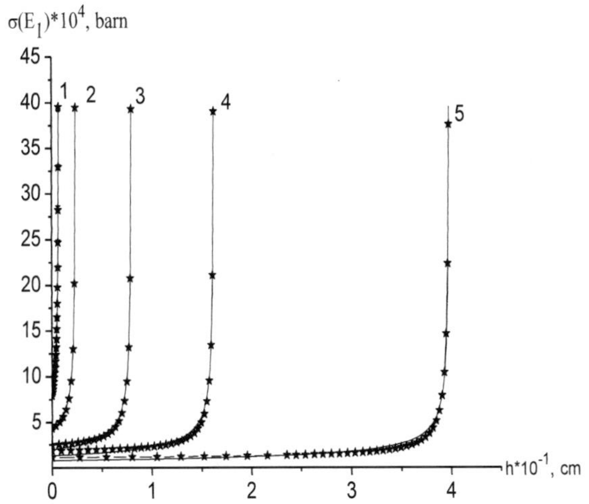


$E_0$ = 5(1), 10(2), 20(3), 30(4), 50(5) MeV.
The points are the calculated data, the solid line is the approximation (curves 1 – 5, respectively)

**Figure 1.** Dependences of the modified CPF cross section for protons in copper from h

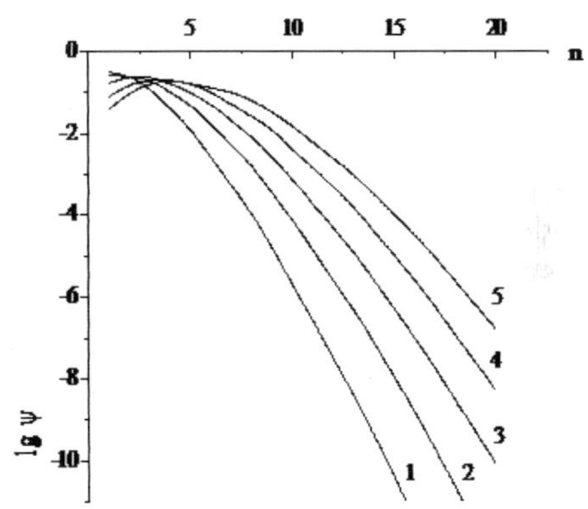


$E_0$ = 10 MeV,
h = 40; 60; 80; 100; 120 μm (1-5)

**Figure 2.** Dependence of CPF on the number of interactions for protons in Cu

The coefficients $\sigma_0$, a, $E_0'$, $k$ are found by the method of least squares, where $E_0'$ is not the initial energy of the primary particle, but is the approximation coefficient. The theoretical correlation ratio for all elements and energies in the range from 5 to 50 MeV ranges from 0.99 to 0.999, which indicates a very good selection of approximations. When choosing approximation curves for protons according to formula (1), difficulties arise that involve a specific analysis of each curve. The choice of optimal values depends on the initial data, for protons we have 4 approximation parameters. In addition, there are situations when the step in h fluctuates strongly, it is small, then large, and then the approximation formula is even harder to pick up. In this case, you need to use techniques to eliminate emerging features. In the region of small initial particle energies for protons, the process of selecting the type of approximation curves is also complicated by the fact that the step in h and the depths themselves are small, while the values of the cross sections are large. Compared with light, for heavy elements, for example Mo, the approximation curve is chosen somewhat easier. With increasing initial energy, the selection of curves is better.

IOP Conf. Series: Materials Science and Engineering **510** (2019) 012024 doi:10.1088/1757-899X/510/1/012024

Calculation of CPFs taking into account the energy losses for protons was carried out according to formula (2). The results of the calculations are shown in Figure 2. Depending on n, for different h, the CPFs first increase, then reach a maximum and then decrease.

The calculation of PKA spectra was carried out according to the formula (3). The results of the calculations are shown in Figures 3 – 5.

The results of the calculations show that the dependence of PKA spectra on PKA energy is a decreasing function. It can be seen that as the initial proton energy increases (Figure 3), the values of the PKA concentration decrease, with increasing penetration depths increase (Figure 4).

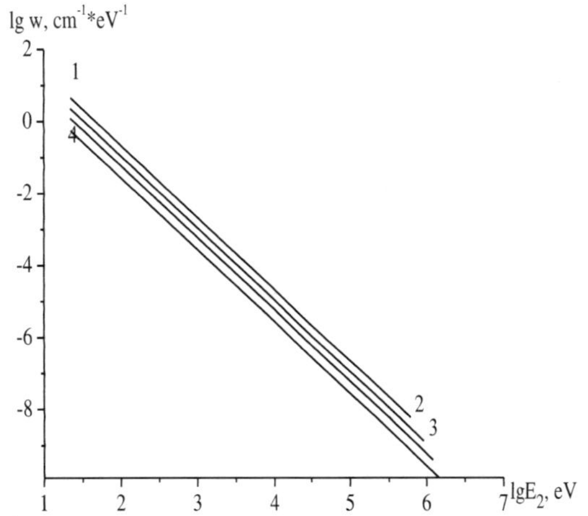


$E_0$ = 10(1); 15(2); 20(3); 30(4) MeV

**Figure 3.** Dependence of w ($E_0$, $E_2$, h) on $E_2$ for protons in Cu at a depth of h = 10 μm

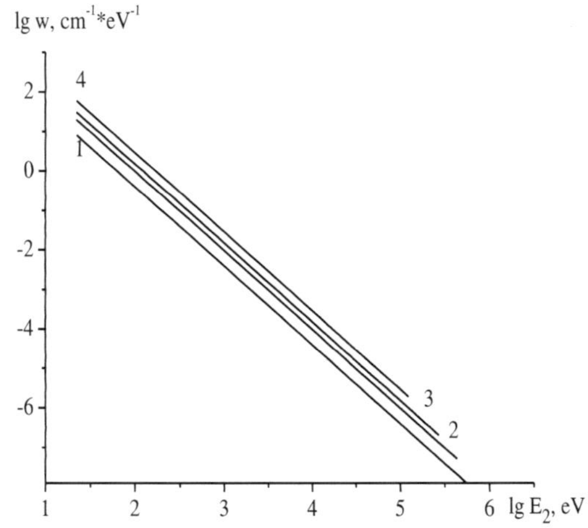


$E_0$ = 10 MeV; h = 20(1), 110(2), 181(3), 226(4) μm

**Figure 4.** Dependence of w ($E_0$, $E_2$, h) on $E_2$ for protons in Cu

The results of PKA spectra calculated using formula (3) are compared with the results obtained earlier in [14] and shown in Figure 5 with a dashed line. We see a good agreement. The small difference is due to the fact that in [14] the average value of the mean free path length for the formation of PKA was taken to calculate PKA spectrum, and in this paper we used a range that takes into account the energy loss for ionization and excitation during PKA generation. To compare the results of calculations, in addition to protons in copper and molybdenum, alpha particles were taken in molybdenum, presented in Figure 5. There is a slight difference. Calculation of the concentration of radiation defects during proton irradiation was carried out according to formula (4). The results are shown in Figure 6. For light targets, the curves increase, reaching a maximum, then decrease to zero. With an increase in the initial energy of the particle, the values of the concentration of radiation clusters increase, the maximum is more pronounced. With increasing atomic number of the target, the behavior is similar. With an increase in the threshold energy $E_c$, the values of the concentration decrease, and the curves pass much lower, the transition through the maximum is made smooth.

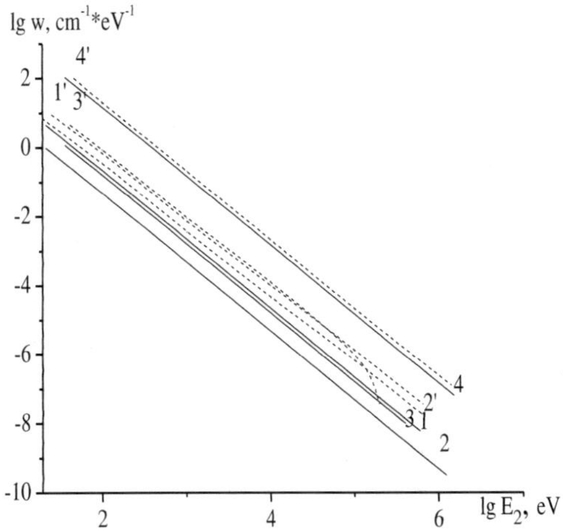


1 – p-Cu, h = 10 μm, $E_0$ = 10 MeV; 2 – p-Cu, h = 8 μm $E_0$ = 20 MeV; 3 – p-Mo, h = 10 μm, $E_0$ = 10 MeV; 4 – α-Mo, h = 10 μm, $E_0$ = 10 MeV (W, $cm^{-1}$, $eV^{-1}$; $E_2$, eV) (calculations have been made using the formula (3), 1′ – 4′ (calculations from [14])

**Figure 5.** PKA spectra at various depths in copper and molybdenum at proton and alpha-irradiation

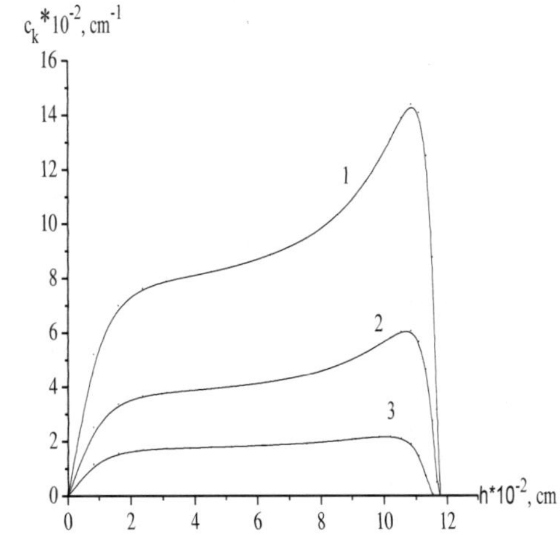


$E_0$ = 25 MeV $E_c$ = 50 keV (1), $E_c$ = 100 keV (2), $E_c$ = 200 keV (3)

**Figure 6.** Dependence of the concentration of radiation defects on the depth at proton irradiation for Cu

## 4. Conclusions

The energy spectra of PKA and the concentration of vacancy clusters in materials irradiated with protons were simulated. The approximation of the expression was performed and the approximation parameters for various values of the initial energy were found. Calculations of cascade-probability functions were performed. Algorithms are presented and calculations of the spectra of primary knocked-on atoms were performed as a function of PKA energy for different values of the initial energy and depths. The results of calculations were compared with calculations made earlier. There is a good agreement. Calculations of the concentration of defects at various values of the threshold energy were made. $C_k$ as a function of h, first it increases, then, reaches a maximum and then decreases.

## Acknowledgments

The study was supported by the Grant of the Ministry of Education and Science of the Republic of Kazakhstan.

## References

[1] Dins J and Vinyard J 1960 *Radiation effects in solid bodies* (M.: Foreign literature)
[2] Damask A and Dins J 1966 *Dot defects in metals* (M.: World)
[3] Kodanova S K, Ramazanov T S, Issanova M K, Golyatina R I and Maiorov S A 2018 Calculation of ion stopping in dense plasma by the Monte-Carlo method *Journal of Physics: Conference Series.* **946** 1 148-154
[4] Smyrnova K V, Pogrebnjak A D, Beresnev V M, Kravchenko Y O and Bondar O V 2018 Microstructure and Physical–Mechanical Properties of (TiAlSiY)N Nanostructured Coatings Under Different Energy Conditions *Metals and Materials International* **24** 2 241-439

[5] Komarov F F 2017 Nano- and microstructuring of solids by swift heavy ions *Physics-Uspekhi.* **187** 5 465-504

[6] Komarov F F, Zayats G M, Komarov AF, Michailov V V and Komsta H 2015 Simulation of radiation effects in SiO2/Si Structures *Acta Physica Polonica A.* **3** 1 854-860

[7] Komarov F, Uglov V, Vlasukova L, Zlotski S and Yuvchenko V 2015 Drastic structure changes in pre-damaged GaAs crystals irradiated with swift heavy ions *Nuclear Instruments and Methods in Physics Research, Section B: Beam Interactions with Materials and Atoms* **360** 4 150-155

[8] Boos E G, Kupchishin A A, Kupchishin A I, Shmygalev E V and Shmygaleva T A *2015 Cascade and probabilistic method, solution of radiation and physical tasks, Boltzmann's equations. Communication with Markov's chains* (Almaty: KazNPU after Abay, Scientific Research Institute for New Chemical Technologies and Materials under al-Farabi KazNU)

[9] Kupchishin A I, Shmygalev E V, Shmygaleva T A and Jorabayev A B 2015 Relationship between Markov Chains and Radiation Defect Formation Processes by Ion Irradiation *Modern Applied Science* **9** 3 59-70

[10] Kupchishin A I, Kupchishin A A, Shmygalev E V, Shmygaleva T A and Tlebaev K B 2014 Calculation of the spatial distribution of defects and cascade-probability functions in the materials *Journal of Physics* 552 1-7

[11] Voronova N A, Kupchishin A I, Kupchishin A A, Kuatbayeva A A and Shmygaleva T A 2018 Computer Modeling of Depth Distribution of Vacancy Nanoclusters in Ion-Irradiated Materials *Key Engineering Materials* **769** 358-363 doi:10.4028/www.scientific.net/KEM.769.358

[12] Deans J and Vinyard J 1960 *Radiation defects in solids* (M.: Inostrannaya literature)

[13] Nemets O F and Gofman Yu V 1975 *Handbook of Nuclear Physics* (Kiev: Naukova dumka)

[14] Boss E G and Kupchishin A I 1988 *Solution of physical problems by cascade-probabilistic method* (Alma-Ata: Nauka)

IOP Conf. Series: Materials Science and Engineering **510** (2019) 012025 doi:10.1088/1757-899X/510/1/012025

# Study of the influence of the electron irradiation dose on the deformation of mylar films taking into account the processes of destruction and crosslinking

**A I Kupchishin[1,2], B G Taipova[1], V M Lisitsyn[3] and M N Niyazov [1]**
[1] Kazakh National Pedagogical University named after Abai, Dostyk 13, Almaty, Kazakhstan
[2] Kazakh National University named after al-Farabi, al-Farabi av 71, Almaty, Kazakhstan
[3] Tomsk National Research Polytechnical University, Tomsk, Russia

E-mail: ankupchishin@mail.ru

**Abstract.** Experimental studies of the dependence of the deformation ε on the dose of electron irradiation D taking into account the processes of destruction and crosslinking for films of the Mylar type have been carried out. It was obtained that the dependence of ε on D for the process of cross-linking is described by a linear and destruction by an exponential function.

## 1. Introduction

The study of the processes occurring in polymers under the influence of various factors makes it possible to predict their behavior in the fields of intense loads. The physico-mechanical properties of the material are determined by the number of defects contained in it and the probability of the occurrence of various reconstructions with their participation [1, 2]. One of the important directions of research in this case is the study of the change in the physicomechanical characteristics of high molecular compounds under the action of high-energy particles, which cause significant transformations in the structure [2]. The mechanism of influence of high-energy particles on high-molecular compounds in the condensed state is characterized by the presence of physical and chemical stages. There is a spatial transformation of matter, the molecular weight of the macromolecule changes, and accumulation and disappearance of various types of chemical bonds, etc., which lead to a variety of properties of the polymer material [3, 4]. The most significant structural changes in the properties of polymers are due to irreversible processes of structuring and destruction. These processes are usually carried out simultaneously, and the predominance of one of them depends on the structure of the material, the nature of the substances present in the system and the irradiation conditions [5]. Structuring improves a number of properties of polymers. The transformation of a linear polymer into a spatial polymer is accompanied by a significant increase in the molecular weight, which leads to its insolubility in organic solvents, to an increase in the mechanical strength of the material, and so on. The destruction of the polymer material reduces its molecular weight. The mechanical strength and the softening temperature during degradation are reduced, since the forces of intermolecular bonds in the polymer are higher, the larger the size of the molecules. Therefore, for practice, it is of great importance to study methods of protecting polymers from degradation [6]. At present, there is a large number of theoretical and experimental studies describing the various stages of the radiation-physical processes taking place in polymer materials, a significant amount of data on radiation-chemical transformations in these media under the action of ionizing radiation under various conditions has been accumulated. It becomes possible to solve not only the problems of predicting the behavior of structural polymers in the fields of ionizing radiation, but also the purposeful creation of materials with predetermined properties [7, 8].

## 2. Experimental

The object of the study is an industrial polyethylene terephthalate film manufactured in the USA (Maylar type) in the form of strips 5 mm wide, working length is 50 mm, 100 μm thick. Irradiation of the samples was carried out in air in special holders on an ELU-6 electron accelerator at 20 °C with an energy of 4 MeV, a current density of 0.5 μA/s, a pulse duration of 5 μs at a repetition rate of 200 Hz. Absorbed doses (D) were $0 - 3 \cdot 10^5$ kGy. Investigation of mechanical characteristics was carried out on a computerized tensile machine of the RMU-0.05-1 type with a clamping speed of 36.09 ± 0.05 mm/min. The movement of the grip associated with the meter was 0.1 mm. The tests were carried out in the form of uniaxial tension using a special reverser at a constant load and the temperature of 20 ± 2°C, relative humidity was (45 ± 5) %. Deformation was determined depending on the dose of irradiation. The obtained experimental data were transmitted directly to Excel tables. Their parameters were set by the user. All the experimental data obtained were mathematically processed (Mathematica 5 system).

## 3. Discussion of the results

Figure 1 (a, b) shows the experimental (points) dependences of the relative elongation (ε) on the dose of electron irradiation D at the temperatures of a) T = 293 K and b) T = 473 K. With an increase of D up to $10^4$ Gy at the temperatures of 293 and 473 K, the relative elongation first slowly increases and at D ~ $10^6$ Gy ε drops sharply to 0. It can be seen from the figure that up to $10^4$ Gy at the temperatures of 293 and 473 K, the crosslinking processes predominate for the Mylar film, and a further increase in the dose leads to destruction of the material (curves 1 and 2). With height of temperature of ε grows. For an interval of doses of $0 - 2*10^4$ Gy the specific elongation at T = 473 K is 1.93 times larger than at T = 293 K.

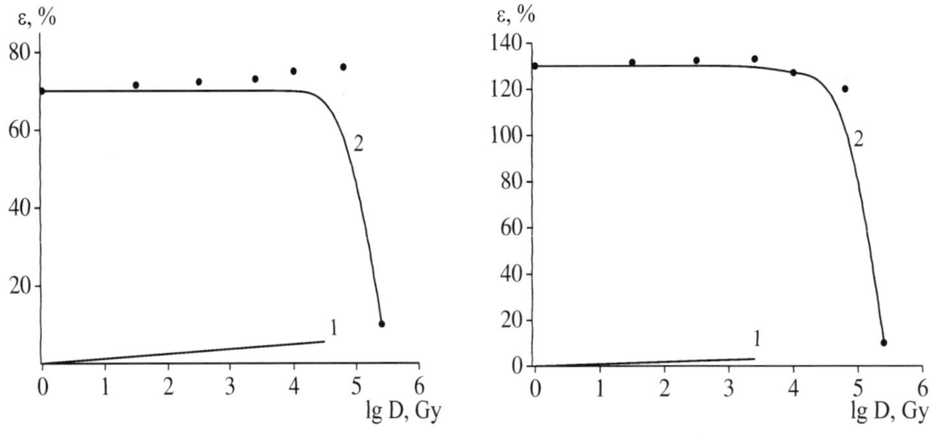


a – 293; b – 473 K
points – experiment; 1 – crosslinking component; 2 – destruction component

**Figure 1**. Dependence of the deformation on the dose of electron irradiation at different temperatures of Mylar type films

We have proposed a generalized model that simultaneously describes the processes of crosslinking and destruction for the whole range of doses from 0 to $10^6$ Gy (before destruction).

The general formula for calculation is:

$$\varepsilon = \varepsilon_1 + \varepsilon_2,$$

$$\varepsilon_1 = \varepsilon_0 \exp\left(-\frac{D}{D_0}\right); \quad \varepsilon_2 = aD,$$

where $\varepsilon_0$ is the maximum value of material deformation before irradiation; $a$ is a coefficient characterizing the crosslinking. Moreover, the crosslinking is taken into account as a linear, and the destruction as an exponential function (Figure 1a and b, curves 1 and 2). For the Mylar-type film, the crosslinking component is negligible and is ~ 5 %. Irradiation of polymers affects the crosslinking and degradation processes, forming chemically unsaturated bonds and gaseous products. This breaks the chemical bonds of C-C in the main and side chains of the macromolecule and the bonds of C-O, C-H, C-N, C-F, C-Cl, Si-C, and C-S. The investigated sample of the Mylar-type film has a complex heterogeneous structure of amorphous regions with a significant proportion of interfibrillar amorphous clusters separating adjacent microfibrils, along with intrafibrillar amorphous interlayers [9, 10]. When the material is irradiated, the absorbed energy causes simultaneous discontinuities of mainly C-C and C-H bonds. Crosslinking polymers are characterized by the recombination of a significant part of the broken C-C bonds and, as a result, the discontinuities of the C-H bonds predominate.

The impact of external loads on the polymer material creates a strain that propagates at a certain rate from the point of application of force from the clamps along the specimen. The weakened fragment of the film closest to the place of application of the load will be the point of discontinuity. The process of destruction consists of two stages: the first stage is the emergence and growth of primary cracks; in the second stage, cracks develop rapidly and germinate through the entire cross section of the sample, leading to its destruction [6]. In the polymer, the cause of the appearance of primary cracks is thermal fluctuations, which lead to a sharp increase in the kinetic energy of individual atoms, and consequently, the chemical bond breaks down in the polymer backbone. Increased temperature and applied stress increases the probability of rupture, especially those bonds that are overstrained [7]. For highly elastic polymers, along with the breakage of chemical bonds, slipping of one part of the sample from the other, accompanied by rupture of intermolecular bonds, takes place during destruction [1, 3].

We note that an increase in temperature to 473 K leads to an increase in strain by 90 %, but the percentage of crosslinking processes decreases. This is confirmed by the curves of Figures la and b. At very high doses, fragments of the installation loosely coupled among themselves elements of the temperature aggravated by the action remain. This is due to the weak lateral branching of the PET polymer chain. At the same time, the number of passing molecules increases and the mechanism of their sweep from the cross-linked systems is violated. For destructive polymers, rapid recombination of broken ends of the chain is difficult, in them radicals are stabilized by the disproportionation reaction, forming two stable end groups, which fix the act of destruction. The competing processes of crosslinking and destruction are ultimately determined by the rate of recombination of the bond rupture.

It was established in [11 – 13] that irradiation in PET reduces the molecular weight by a factor of 3 compared to the unirradiated samples. From elemental analysis, it was also found that with increasing absorbed dose, the carbon content (n) decreases by 2.60 % and hydrogen by 0.18 %, but at the same time the oxygen content increases by 2.78 %. All these data indicate the discontinuities of macromolecules and the predominance of the process of destruction. Studies have shown a reduction in the optical density of the film with an increase in the radiation dose, which also indicates restructuring processes involving the displacement or misorientation of optical centers, the intensification of the destruction of polymer chains, and gas formation in the bulk of the polymer under the action of fast electrons.

## 4. Conclusions

1. Under the influence of temperature, irradiation and static loading on Mylar-type films, crosslinking and destruction processes are intensified.

2. Prior to doses of $10^4$ Gy at the temperatures of 293 and 473 K, the predominance of crosslinking processes is observed. A further increase in the irradiation dose to $10^6$ Gy increases the probability of rupture of chemical and intermolecular bonds associated with the slip of one part of the sample from the other, accompanied by a rupture, with a weak lateral branching of the PET polymer chain. At the same time, the mechanism of the development of molecules from cross-linked systems is violated.

3. As the temperature of the irradiated material increases, the deformation increases greatly, but the cross-linking process decreases, which is due to the instability of the fragments of the mounting loosely coupled elements, which are aggravated by the effect of temperature. Thermal fluctuations lead to a sharp increase in the kinetic energy of individual atoms, and as a result, the chemical bond breaks down in the polymer backbone.

4. A generalizing model of deformation dependence on the dose of electron irradiation is proposed, which takes into account competing processes of cross-linking and destruction. It is obtained that the process of cross-linking is described by a linear function, and the destruction process – by exponential one.

## Acknowledgments

The study was supported by the Grant of the Ministry of Education and Science of the Republic of Kazakhstan.

## References

[1] Zuev V V, Uspenskaya M V and Olekhnovich A O 2010 *Physics and chemistry of polymers Textbook* (SPb.: SPbSU ITMO)

[2] Kudaykulova S K, Iskakov R M, Kravtsova V D, Umerzakova M B, Abadie M, Zhubanov B A at al 2006 *Polymers for special purposes* (Almaty)

[3] Ed. I G Assovskiy and others 2011 *Chemical and radiation physics* (M.: TOPUR PRESS)

[4] Ivchenko V A 2016 Atomic structure of cascades of atomic displacements in metals and alloys after different types of radiation *IOP Conf. Series: Materials Science and Engineering* 110012003 1-5

[5] Kudashev S V, Urmantsev U R, Seleznev G V, Rakhimov N A and Zheltobryukhov V F 2012 Surface modification of PET granules by fluoride-containing urethanes *Journal of Applied Chemistry* **85** 11 1860-66

[6] Shah V 2009 *Reference guide for testing plastics and analysis of the causes of their destruction trans. from English* (Ed. AYa Malkin, SPb.: Nauchnyye osnovy i tekhnologii)

[7] Ardekani S M, Dehghani A, Al Maadeed M A, Wahit M U and Hassan A 2014 Mechanical and thermal properties of recycled poly(ethyleneterephthalate) reinforced newspaper fiber composites, *Fibers Polym.* 15 1531-38

[8] Rouhi S, Alizadeh Y and Ansari R 2014 On the interfacial characteristics of polyethylene/single-walled carbon nanotubes using molecular dynamics simulations *Applied Surface Science* **292** 958-970

[9] Gerasimenko N N, Voronova N A, Kovtunets V A, Taipova B G, Kupchishin A I and Omarbekova Z A 2003 Study of the mechanical properties of polymer films effecting on their by static load and temperature *Proceedings of the universities. Materials of electronic technology* 6-12

[10] Kupchishin A I, Taipova B G, Kupchishin A A, Voronova N A, Kirdyashkin V I and Fursa T V 2016 Catastrophic models of materials destruction *IOP Conf. Series: Material Science and Engineering 110 012037* 1-5
[11] Kupchishin A I, Taipova B G, Kupchishin A A and Kozhamkulov B A 2015 Study on the physical and mechanical properties of composites based on polyimide and polycarbonate *Mechanics of composite materials* **51** 1 115-118
[12] Kozhamkulov B A, Kupchishin A I, Bitibaeva Zh M and Tamuzhs V P 2017 Radiation defect formation and destruction of composite materials under electron irradiation *Mechanics of composite materials* **53** 1 89-98
[13] Kupchishin A I, Taipova B G and Voronova N A 2016 Study of the influence of filler on the Structure Treatment systems biomaterials *IOP Conf. Series: Materials Science and Engineering: 168012015* 1-4

# Study of the influence of the electron and gamma radiations on the bottom sediments of Sorbulak lake

**A I Kupchishin[1,2], M N Niyazov[1], N N Gerasimenko[3] and N A Voronova [1]**
[1] Kazakh National Pedagogical University named after Abai, Dostyk 13, Almaty, Kazakhstan
[2] Kazakh National University named after al-Farabi, al-Farabi av 71, Almaty, Kazakhstan
[3] National Research University Moscow Institute of Electronic Technology, Moscow, Russia

E-mail: ankupchishin@mail.ru

**Abstract.** The influence of electron and gamma radiation on the bottom sediments of Sorbulak Lake was investigated. It is shown that irradiation with both electrons and gamma quanta changes significantly the dependence of intensity (I) on time (t). From the curves of intensity versus time, regularities were obtained which have the same character both for the unirradiated and for the irradiated biomaterial. Curve dependencies I (t) are satisfactorily described within the framework of the exponential model.

## 1. Introduction

The protection of the environment and the preservation of sustainable development of industry is an important environmental problem. Much attention is paid to the study of bottom sediments accumulation conditions in lakes, as a factor that depends objectively on the state of the reservoir [1]. Expanding the use of natural resources, introducing new and modernizing existing technologies and increasing production lead to the ecological problem of water purity, bottom sediments (BS) and the use of coastal materials in the national economy [2, 3]. The process of accumulation of heavy metals in bottom sediments of inland water bodies, rivers and canals is characterized by a rather high content of phosphate ions, which bind heavy metals in hardly soluble compounds, leading to their transition to BS [4, 5]. Various methods are used on local sewage treatment plants for the extraction of heavy metals from wastewater: precipitation with alkaline reagents, adsorption, ion exchange, and sometimes biological purification. In treated wastewater, the heavy metal content does not reach zero using common methods of treatment. Uncleaned water is supplied to the sewers, to municipal wastewater treatment plants. Some of them transfers to the natural reservoirs, where also contaminated rain water flows, which leads to contamination of natural reservoirs with BS [6, 7]. The degree of purification of bottom sediments and biomaterials is largely determined by the rate of their sedimentation and is described by the silt index, as microorganisms participate in the natural purification of reservoirs. Well-deposited materials have an index of about 60 mL/g, a less dense 80-90 mL/g, and an index above 300 mL/g indicates a disruption of the treatment plant [8]. If you use the method of biological post-treatment for reservoirs, in addition to the above, the indicators adopted in sewage treatment practice (biochemical oxygen consumption, nitrate and ammonium nitrogen, nitrite content), as well as the characteristic of activated sludge, are determined additionally. Determination of their quantity by dry weight, ash content, and observation of the vital activity of protozoa and rotifers by means of a microscope are conducted to obtain the state of microorganisms after irradiation with particle flows [9, 10]. The present work is devoted to the study of the influence of electron and gamma radiation on the bottom sediments of Sorbulak Lake.

## 2. Experimental

To study non-irradiated and irradiated BS we developed a special installation. The circuit is shown in Figure 1. It consists of a laser source of red color with a length of a monochromatic wave of 780 nm; radiation detector which is a photodiode sensor with a measurement range of 0 – 5000 lux; electronic unit of "Science cube" is a new digital measuring laboratory; racks and various holders.

Bottom sediments were placed in a flask with 25 mL of water, thoroughly shaken and installed between the laser and the detector. Study of the properties of bottom sediments depends on the concentration, so its determination is carried out at a constant dose of BS equal to 30 g/l. If the mixture to be analyzed has a smaller dose, then its thickening is necessary, and if the dose is greater than this, dilution occurs. In studies conducted, the dependence of light intensity on time was taken. The time of the experiment is approximately two minutes.

Irradiation of samples with electrons and photons was carried out on a linear accelerator ELU-6 with an energy of 2 MeV in the air. Samples for irradiation were placed at a distance of 40, 60 cm from the exit window of the accelerator. The magnitude of the beam current was 0.1 – 0.2 $\mu A/cm^2$, the dose of irradiation by electrons was 500 kGy, and the photons by 3P. The irradiation time was 77 minutes and 48 seconds, respectively. Also, studies were carried out on unirradiated and irradiated samples obtained two years ago in similar doses.

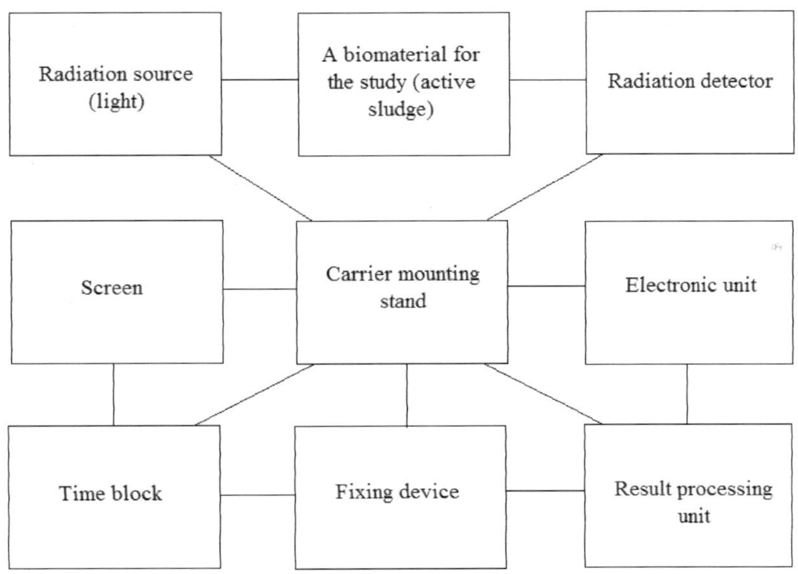


**Figure 1.** Scheme of the experiment on the passage of radiation through a biomaterial

## 3. Results and discussion

Experimental studies have been carried out on the effect of electron and gamma radiation on bottom sediments by the optical method. It was obtained that the complete subsidence occurs within two minutes. Figures 2 and 3 show the time dependences of the intensity for different types of radiation. As follows from these figures, the intensity increases with time, first in the time interval 0 – 20 seconds - increases sharply, and then gradually goes to saturation. Experiments carried out on bottom sediments irradiated with electrons and photons (gamma rays and X-rays) indicate that particle beams significantly influence the specific dependences. In this case, the nature of the dependence of I on t after irradiation does not change. When the biomaterial is irradiated by electrons with 500 kGy dose, the sedimentation rate increases in comparison with the unirradiated samples. The quality of treatment systems and natural purification can be judged from the sedimentation rate of bottom sediments, which affect the value of the silt index. In Figure 3, for curve 2, biomaterial irradiated with photons 3 P, the sedimentation rate decreases. It follows from this that the activity of microorganisms in the bottom sediments is activated by irradiation. After conducting a series of experiments and calculations, it was found that the experimental error is ~ 5 %.

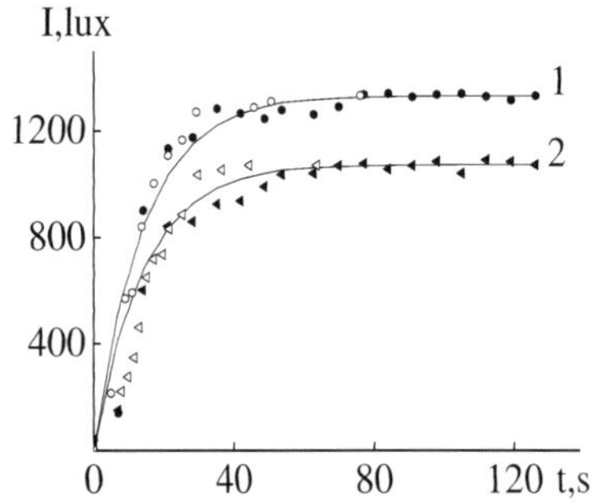


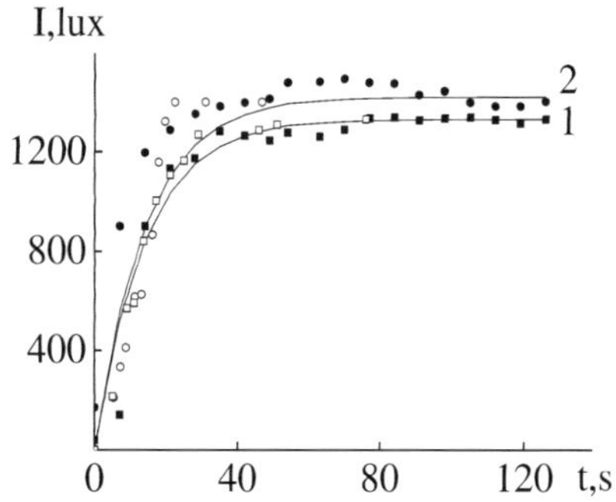


light figures – experimental data of 2016; black figures – experimental data of 2018; solid line – calculation

**Figure 2**. The intensity versus time for unirradiated bottom sediments (1) and those irradiated by 500 kGy electrons (2)

**Figure 3.** The intensity versus time for unirradiated bottom sediments (1) and those irradiated by photon of 3P dose

Comparing the experimental data of the samples irradiated in 2016 and 2018, it can be said that the change in the behavior of microorganisms after irradiation with a high-energy electron flux has been insignificant for different years. This is due to their extermination after irradiation by electrons with a large dose for a long time (500 kGy, 77 minutes). The flux of photons of gamma radiation, in turn, increases the activity of the biomaterial immediately after its irradiation (according to experiments conducted in 2016). The material irradiated in 2018 is less active. The reason for this behavior can be a particular extinction of microorganisms for two years in containers. The experimental data are explained within the framework of the exponential (cascade-probabilistic) model proposed by the authors [11]:

$$I = I_0\left(1 - \exp\left(-t/t_0\right)\right),$$

where $I_0$ is the intensity at complete sedimentation of bottom sediments, and $t_0$ is the time during which the $I/I_0 - 1$ material decreases by *e* times. For this study $t_0$ = 12 seconds. As follows from the figure, the calculated dependences satisfactorily describe the experimental curves obtained.

**Table 1.** Dependence of parameters $I_0$, $t_0$ on dose and type of irradiation

| D | $I_{max}$, (Lux) | $I_0$, (Lux) | $t_0$, (s) | Year |
|---|---|---|---|---|
| 0 | 1330 | 1330 | 14 | 2018 |
| 500 (kGy) | 1072 | 1070 | 12 | 2016 |
| 500 (kGy) | 1072 | 1070 | 12 | 2018 |
| 3 (R) | 1420 | 1420 | 15 | 2016 |

IOP Conf. Series: Materials Science and Engineering **510** (2019) 012026 doi:10.1088/1757-899X/510/1/012026

| | | | | |
|---|---|---|---|---|
| 3 (R) | 1420 | 1400 | 15 | 2018 |

## 4. Conclusions

1. The influence of electron and gamma radiation on the bottom sediments of Lake Sorbulak was studied. From the intensity versus time curves, regularities have been obtained that are of the same nature for both unirradiated and irradiated biomaterials. In the time interval 0 – 20 seconds, the intensity increases sharply, and at t > 20 s gradually goes to saturation.
2. Both electron and gamma irradiation of biomaterial samples leads to a significant change in the optical properties of bottom sediments. The reason for this is a reduction in the activity of microorganisms.
3. The curves of the dependence of I on t are described satisfactorily by the exponential model.

**Acknowledgments**
The study was supported by the Grant of the Ministry of Education and Science of the Republic of Kazakhstan.

## References

[1] Tyumenev S D and Dostay Zh D 2009 *Water supply and sanitation management of Almaty city* (Almaty: K.I. Satpayev Kazakh National Technical University. Institute of Geography)
[2] Guschina L A and Harwood J L 2013 Algal lipids and theirmetabolism *Algae for biofuels and energy, Eds. M.A. Borowitzka, N.R. Moheimani. Dordrecht; Heidelberg: New York; London: Springer* 17-36
[3] Feinendegen L E 2016 Quantification of Adaptive Protection Following Low-dose Irradiation *Health Phys.* **110 3** 276-280
[4] Solovchenko A Ye, Lukyanov A A, Vassilyeva S G, Savinina Ya V, Solovchenko O V and Lobakova Ye S 2013 Possibilities of biotechnological processing of agricultural wastes with the use of microalgae *Bulletin of Moscow University. Ser. 16. Biology* **4** 38-48
[5] Solovchenko A, Khozin-Goldberg I, Recht L and Bouss ba S 2011 Stress-induced changes in optical properties, pigmei and fatty acid content of Nannochloropsis sp.: implications f(non-destructive assay of total fatty acids *Mar. Biotechnc.* **13** 527-535
[6] Sivakumar G, Xu J, Thompson R W, Yang Y, Randol Smirh P and Weathers P J 2011 Integrated green algal technology for bio re mediation and biofuel *Bioresour. Technol.*
[7] Kirienko N I, Ussenko O M and Mussiy T O 2014 Variability of the biochemical composition of algae (review) *Gidrobiological Journal* **50 5** 54-71
[8] Pittman J K, Dean A P and Osundeko O 2011 The potential of sustainable algal biofuel production using wastewater resources *Bioresour. Technol.* **102** 17-25
[9] Dzyuban A N 2014 Methanogenesis and destruction of organic matter in bottom sediments of technogenic water bodies *Water Resources* **41 3** 330-338
[10] Braun R, Brachtl E and Grasmug M 2003 Co digestion of protein aceous industrial waste *Appl. Biochem. Biotechnol.* **109** 139-153
[11] Voronova N A, Kupchishin A I and Taipova B G 2018 Nanoclusters and Electron Irradiation Effect on Mechanical Properties of Polyimide-Based Composite *Key Engineering Materials* **769** 72-77
[12] Voronova N A, Kupchishin A I, Niyazov M N and Lisitsyn V M 2018 Uniaxial Stress and Electron Irradiation Effects on Nanochains Straightening in Film Polymer Materials *Key Engineering Materials* **769** 78-83

# Holographic method of the nanoparticles diagnostics in a fluid

**V I Ivanov, O O Ovseychook and A V Myagotin**

Department of Physics and Theoretical Mechanics, Far Eastern State Transport University, Khabarovsk, Serysheva street, 47, Russia

E-mail: magistrfestu@yandex.ru

**Abstract**. The dynamic holography method is analyzed for the diagnostics of nanoparticles in a fluid. The concentration nonlinearities caused by electrostriction and thermodiffusion effects as well as the reorientational cubic response in nanosuspensions are considered. The results are relevant to the study of the optical diagnostics of such materials.

## 1. Introduction

Multicomponent nanoparticles medium (nanosuspensions, microemulsions, aerosols) are characterized by a series of specific mechanisms nonlinearities that not appear in single-component media [1-3]. In particular, these include the concentration non-linearity caused by the redistribution of the dispersed component in a two-phase medium in a laser field. The concentration flows in the medium can be caused by different mechanisms of interaction of radiation with matter. The optical nonlinear response of the nanosuspensions includes the parameters of the nanoparticles. The thermodiffusion effect (Soret effect [4]) in the binary mixtures was experimentally investigated by the optical nonlinear methods in many works [5-8]. Another mechanism of optical nonlinearity of the medium is due to the forces operating on the particles of the dispersed phase in gradient light field [9-12].

The purpose of this work is the theoretical analysis of the contribution of concentration and orientation nonlinear mechanisms in the formation of cubic nonlinear response in the nanosuspension.

## 2. Electrostrictive mechanisms of the cubic nonlinearity in nanosuspension

The coefficient of cubic nonlinearity for the concentration mechanism is determined by the concentration nonlinearity:

$$n_2^{eff} = (\mathrm{d}n/dC)(dC/dI) \tag{1}$$

Where $n$ the refraction index of the medium is, $I$ is the radiation intensity, $C$ is the concentration of nanoparticles, $n_2^{eff} = (dn/dI)$ is the coefficient of effective cubic nonlinearity.

In nanosuspension the particle radius is much smaller than the radiation wavelength $\lambda$, therefore the refractive index of the medium is proportional to the concentration of particles [3]:

$$n = n_1(1 + f\varepsilon), \tag{2}$$

where $\varepsilon = (n_2 - n_1)/n_1$ ; $n_1$ and $n_2$ are the refractive indices of the liquid and the dispersed phase, respectively; $f = v_0 C$ is the volume fraction of the dispersed phase, $a_0$ is the radius of the nanoparticle, $v_0 = (4/3)\pi a_0^3$ is the volume of the nanoparticle.

The system of balance equations for the concentration of particles and heat flow for this case is written as follows [6]:

$$dC / dt = -\operatorname{div}(J_d + J_{el}). \quad (3)$$

Here $(J_d + J_{el})$ are concentration flows:

$$J_d = -D\nabla C, \quad (4)$$

$$J_{el} = \gamma C\nabla I, \quad (5)$$

where $J_d$ is the diffusion flow, $D$ is the diffusion coefficient of particles, $J_{el}$ is the electrostrictive flow, $\gamma = (2\beta b / cn_1)$, ($\beta$ and $b$ are polarizability and mobility of microparticle, respectively), $c$ is the light velocity, $I = I_0(1 + \sin Kx)$ is the intensity spatial distribution along the layer of the medium (x-axis), $K$ is the space wave vector of the elementary hologram.

Assuming a one-dimensional problem, the solution of equations (7) is founded in the form of:

$$C(x,t) = C_0 + C_1(t)\sin Kx, \quad (6)$$

where $C_0$ is the average particle concentration.

The linearized system is easily solved (with the approximation $(C_1/C_0) << 1$):

$$C_1(t) = C_0 I_0 I_{sat}^{-1}[1 - \exp(-t/\tau_d)], \quad (7)$$

$$I_{sat} = D\gamma^{-1}, \quad (8)$$

$$\tau_d = (K^2 D)^{-1}, \quad (9)$$

where $I_{sat}$ is the "saturation" intensity, $\tau_d$ is the diffusion relaxation time.

For great intensity beam ( $I >> I_s$ ) the nonlinear response of the nanosuspension does not match to the cubic nonlinearity, because the concentration is not a linear function of the light intensity.

Using the stationary solution (13) for the amplitude of concentrations grating the expressions for nonlinear coefficients can be obtained [13]:

$$n_2^{str} = n_1 \varepsilon \varphi_0 I_{sat}^{-1}, \quad (10)$$

where $\varphi_0 = (4/3)\pi a_0^3 C_0$ is the initial volume fraction of the dispersed phase.

In opposite to the thermal nonlinearity the electrostriction cubic coefficient $n_2^{str} = n_a \delta\phi_0 I_{sat}^{-1}$ does not depend on the relaxation time.

It was experimentally obtained the value $10^{-6}$ $cm^2/kW$ for electrostrictive nonlinearity coefficient in nanosuspensions (so-called artificial Kerr medium) [11,14].

## 3. Thermodiffusion effect

In two-component fluid the heat flow also can cause concentration stream arising from occurrence of thermodiffusion effect. A thermally induced mechanism of particle drift in a nonuniform temperature field is known as thermal diffusion (thermophoresis) or Soret effect in liquid-phase binary mixtures [5].

The balance equations for the concentration of nanoparticles for this case is written as follows

$$c_P \rho dT / dt = -\mathrm{div} J_1 + \alpha I_0 (1 + Sin Kx), \qquad (11)$$

$$dC / dt = -\mathrm{div}(-D \nabla C - C(1-C) D_T \nabla T), \qquad (12)$$

where $J_1 = -\chi \mathrm{grad} T$ is the heat flow, $T$ is temperature of the material, $\chi$ is the thermal conductivity of the material, $c_p$ and $\rho$ are the specific heat capacity and density of the medium, $\alpha$ is the absorption coefficient of the medium,

The solving of the one-dimensional problem for the temperature amplitude is founded in the next form:

$$T_1 t) = T_1(t) \sin Kx, \qquad (13)$$

The solution (11) for the temperature amplitude $T_1$ is:

$$T_1 = T_1^s (1 - \exp\{-t / \tau_T\}), \qquad (14)$$

$$T_1^s = \alpha I_0 \tau_T (c_P \rho)^{-1}, \qquad (15)$$

$$\tau_T = c_P \rho \chi^{-1} K^{-2}, \qquad (16)$$

where $\tau_T$ is the thermal relaxation time, $T_1^s$ is the temperature amplitude in the steady state regime.

Let's consider an expression for the holographic sensitivity of the dispersion medium, taking into account both (thermodiffusion and electrostriction) concentration mechanisms.

The linearized equation (12) is easily solved by taking into account (15):

$$C_1 = C_1^s (1 - \exp\{-t / \tau_d\}), \qquad (17)$$

$$C_1^s = \alpha \chi^{-1} I_0 S_T K^{-2}, \qquad (18)$$

where $C_1^s$ is the concentration amplitude in the steady state regime.

The expression for nonlinear coefficients for this case is the next:

$$n_2^{\mathrm{thd}} = n_1 \varepsilon \varphi_0 \alpha \chi^{-1} S_T K^{-2}, \qquad (19)$$

It was experimentally obtained the value $10^{-7}$ cm$^2$/kW for nonlinearity coefficient in nanosuspensions [5].

## 4. Orientation nonlinearity

The optically induced birefringence in the nanosuspension with non-spherical particles presents the orientation mechanism of the cubic nonlinearity [1]. In the classical model the macroscopic polarization can be obtained by performing an average on the microellipsoid angular displacement of the ensemble of identical microellipsoidal particles [15]. The induced dipole appears due to light field and the opposite action of the rotational diffusion. The model gives the next expression for cubic nonlinearity coefficient $n_2^{ort}$:

$$n_2^{ort} = 8\pi^2 C(\alpha_{II} - \alpha_\perp)^2 / 15 n_0^2 k_B T, [\mathrm{cm^2/kW}], \qquad (20)$$

where $C$ is the particle concentration, $k_B$ is the Boltzman factor, $\alpha_{II}$ and $\alpha_\perp$ are the values of microellipsoid polarizability along the semi-major and the semi-minor axes $a$ and $b$. The numerical estimation gives the theoretical value about $10^{-10}$ cm$^2$/kW for $n_2^{ort}$.

The values of the particle polarizabilities are given by the expressions [15]:

$$\alpha_{II} = (ab^2/3)\varepsilon_a(\varepsilon_r - 1)/[1 + (ab^2/2)(\varepsilon_r - 1)A_1(a,b)], \qquad (21)$$

$$\alpha_{\perp} = (ab^2/3)\varepsilon_a(\varepsilon_r - 1)/[1 + (ab^2/2)(\varepsilon_r - 1)A_2(a,b)], \tag{22}$$

where $A_1$ and $A_2$ are shape dependent coefficients, $a$ and $b$ are the semi-major and the semi-minor axis length, $\varepsilon_a$ is the fluid dielectric constant, $\varepsilon_r$ is the ratio of dielectric constant of the nanoparticle material to $\varepsilon_a$.

The dynamics of the induced birefringence is described by a Debye-like rotational diffusion equation. In the case of cylindrically symmetric particles one obtains the next solution:

$$\Delta n^{\text{ort}}(t) = \Delta n_{st}^{ort}[1 - \exp(-t/\tau_{or})], \tag{23}$$

where $\Delta n_{st}$ is the steady state value of induced birefringence, $\tau_{or}$ is the Debye's response time, $\eta$ is the fluid viscosity, $G(b/a)$ is a form factor which depends on the $(b/a)$ ratio.

$$\tau_{or} = 8\pi\eta a^3[1 - (b/a)^4]/\{9k_B T[2 - (b/a)^2]G(b/a) - 1\}, \tag{24}$$

This type of nonlinearity has been observed experimentally in nanosuspensions [15]. The measured cubic coefficient $n_2^{ort}$ is $10^{-7}$ $cm^2/kW$. This value was interpreted in the model for the particles with intrinsic material birefringence [15].

## 5. Diffraction efficiency of the dynamic hologram

The dynamic holography method (Forced Raleigh Scattering) based on the recording of the dynamic holograms in nonlinear material while reading it. The next expression defines a diffraction efficiency of the hologram [1]:

$$\eta = I_1/I_0, \tag{25}$$

where $I_0$ is the intensity of the incident beam; $I_1$ is the intensity of the diffracted beam. The diffraction efficiency of a thin phase hologram is[1]:

$$\eta = t_0^2 J_1^2(\varphi_1), \tag{26}$$

where $t_0$ is amplitude transmittance of the unlighting hologram; $\varphi_1$ is amplitude modulation after light exposition; $J_n$ is the Bessel function of the n-th order.

If medium is transparent and the modulation amplitude is small ($\varphi_1 << 1$), the diffraction efficiency is equal:

$$\eta = \left\{2\pi n_2 I_m L\lambda^{-1}\right\}^2, \tag{27}$$

where $L$ is the thickness of a nanofluid layer.

The summarized expression for diffraction efficiency is the next:

$$\eta = \{2\pi L\lambda^{-1}I_0[\alpha\tau_T(c_P\rho)^{-1}(\partial n/\partial T) + n_1\varepsilon\varphi_0(\alpha\chi^{-1}S_T K^{-2} + I_{sat}^{-1}) + 8\pi^2 C(\alpha_{II} - \alpha_{\perp})^2/15n_0^2 k_B T]\}^2. \tag{28}$$

where $(\partial n/\partial T)$ is the thermal refractive index coefficient of the fluid.

This expression includes the thermal expansion (first term), the concentration nonlinearity (second term) and the orientation nonlinearity (third term). All types of the nonlinearities are characterised by the various relaxation times (formulas (16), (9) and (24) accordingly).

## 6. Conclusions

The dynamic holography method for the nanosuspension diagnostics was analyzed taking into account the concentration and orientation nonlinearities. The expression was received for the time dependence of the diffraction efficiency of the dynamic hologram in nanosuspension. The results demonstrate the extensive possibilities of the optical diagnostics of such nano-materials [16-20].

**References**

[1] Shen Y R 1984 *The principles of nonlinear optics* (Wiley, New York) p 112-119
[2] Malacarne L C, Astrath N G C, Medina A N, Herculano L S, Baesso M L, Pedreira P R B, Shen J, Wen Q, Michaelian K H and Fairbridge C 2011 *Optics Express* **19** 4058
[3] Ivanov V I, Ivanova G D, Kirjushina S I and Mjagotin A V 2016 *Journal of Physics: Conference Series* **735** 012013
[4] Morozov K I 2009 *Phys. Rev. E* **79** 031204
[5] Leppla C and Wiegand S 2003 *Philosoph. Mag.* **83** 1989
[6] Ivanov V I, Ivanova G D and Khe V K 2017 *Proc. SPIE* **10176** 1017607
[7] Lee W, El-Ganainy R, Christodoulides D, Dholakia K and Wright E 2009 *Optics Express* **17** 10277
[8] Piazza R 2009 *Soft Matter* **4** 1740–1744
[9] Vicary L 2002 *Philos. Mag.* B. **82** 447-452
[10] Ivanov V, Ivanova G, Okishev K and Khe V 2017 *IOP Conf. Ser.: Mater. Sci. Eng.* **168** 012045
[11] El-Ganainy R, Christodoulides D, Rotschild C and Segev M 2007 *Opt. Express* **15** 10207
[12] Chintamani P, Shalini M, Agnel P, Meera V, Tejas I H and Radha S 2014 *International Journal of Chemical and Physical Sciences* **3**(5) 44
[13] Ovseychook O O, Ivanov V I, Mjagotin A V and Ivanova G D 2018 *IOP Conf. Series: Journal of Physics: Conf. Series* **1038** 012091
[14] Philip J and Nisha M 2010 *Journal of Physics: Conference Series* **214** 012035
[15] Pizzoferrato R, Marinelly M, Zammit U, Scudiery F, Martelluchi S and Romagnoli M 1988 Optics Communications **68** 231.
[16] Kamanina V, Kuzhakov P V, Serov S V, Kukharchik A A, Petlitsyn A A, Barinov O V, Borkovskii M F, Kozhevnikov N M and Kajzar F 2013 *Proc. SPIE* **8622** 86221B
[17] Ivanov V I , Ivanova G D, Okishev K N and Khe V K 2016 *Proc. SPIE* **10035** 100354Y
[18] Ishaaya A A, Vuong L T, Grow T D and Gaeta A L 2008 *Opt. Lett.* **33** 13-15
[19] Mjagotin A V, Ivanov V I and Ivanova G D 2017 *Proc. SPIE* **10176** 101761Z
[20] Polyakov P, Luettmer-Strathmann J and Wiegand S 2006 *Journal of Physical Chemistry B* **110** 26215- 26224

IOP Conf. Series: Materials Science and Engineering **510** (2019) 012028 doi:10.1088/1757-899X/510/1/012028

# Analysis of dynamic responses in a composite poroelastic body using the time-stepping boundary element scheme built on the Lobatto method nodes

**S M Aizikovich[1], A N Petrov[1] and S A Tsvetkov[2]**

[1]Research and Education Center "Materials", Don State Technical University,1 Gagarin Sq., 344010, Rostov-on-Don, Russia

[2]National Research Lobachevsky State University of Nizhny Novgorod, 23 Prospekt Gagarina, 603950, Nizhny Novgorod, Russia

andrey.petrov@mech.unn.ru

**Abstract**. In this paper, the dynamic analysis of piecewise-homogeneous poroelastic bodies on the example of a composite three-dimensional column is presented. Biot's model of the material, consisting of a solid phase and two fluid phases (liquid and gas) that filling the pores, is used to describe the poroelastic medium. The boundary element method is used in combination with the time-step method for the numerical inversion of the Laplace transform. The points in the complex plane for calculating the solution are chosen in a special way using the Lobatto method. The two- and the three-stage schemes are considered.

## 1. Introduction

The research continues a series of the works devoted to the application of the boundary elements method combined with the time-step schemes built on the Runge-Kutta methods nodes in problems of the poroelastic dynamics. The results of applying this approach in problems of the dynamics of homogeneous and composite fully saturated poroelastic bodies for the cases of two-stage schemes Radau and Lobatto, were published previously [1, 2]. As well as for problems of the dynamics of homogeneous partially saturated poroelastic bodies for the cases of two and three-stage schemes Radau and Lobatto [3, 4]. In this paper, the boundary element method is generalized for composite partially saturated poroelastic bodies. Two- and three-stage Lobatto methods are used for the iterative process organization.

In general, the approach is based on the convolution quadrature method (CQM) proposed by Lubich in 1988 for discretization of the convolution integral [5, 6]. The method has enlisted the considerable interest as a technique for applying BEM in dynamic problems (CQ-BEM), in which the classic time-stepping schemes show instability and numerical damping [7-10]. Later in various studies it was shown that the time-stepping BEM scheme, based on the Runge-Kutta methods, provides better accuracy than the scheme based on the linear multi-step method [11-15]. Banjai and Sauter also proposed a formulation for uncoupled problems in the Laplace transform domain, within which the technique actually works as an inverse transform method. The approach posses the original method benefits related to the numerical stability and has the time step value as the only parameter, which is determined by the physical characteristics of the problem.

## 2. Models and methods

A set of fully coupled governing differential equations of a porous medium saturated by two compressible fluids (water and air) subjected to dynamic loads is considered. In this formulation the solid skeleton displacements $u_i$, water pressure $p^w$ and air pressure $p^a$ are presumed to be independent variables [16]. The final differential equations in the Laplace domain yield

$$\begin{bmatrix} B_1\delta_{ij} + B_2\partial_i\partial_j & B_3\partial_i & B_4\partial_i \\ B_5\partial_j & B_6 & B_7 \\ B_8\partial_j & B_9 & B_{10} \end{bmatrix} \begin{bmatrix} \hat{u}_i \\ \hat{p}^w \\ \hat{p}^a \end{bmatrix} = \begin{bmatrix} -\hat{F}_i \\ -\hat{I}^w \\ -\hat{I}^a \end{bmatrix}, \mathbf{x} \in \Omega,\ \Omega \subset \mathbb{R}^3, \quad (1)$$

where $\hat{F}_i$, $\hat{I}^w$, $\hat{I}^a$ are bulk body forces and symbol "^" denotes Laplace transform with complex variable $s$. The expressions for the differential operator coefficients are presented in Appendix A. The parameters of the model of a poroelastic material and their values used in further calculations are listed in table A1. Equation (1), supplemented by boundary conditions, fully describes the boundary value problem in the representations of the 3D isotropic dynamic theory of poroelasticity. When a problem involves more than two poroelastic bodies in full contact, equation (1) is written for each body in partitioned form and the resulting equations are coupled together through equilibrium and compatibility at their interface [17].

The boundary-element technique is based on the use of a regularized boundary integral equation (BIE) direct approach:

$$\int_\Gamma \left(\mathbf{T}(\mathbf{x},\mathbf{y},s)\mathbf{u}(\mathbf{x},s) - \mathbf{T}^0(\mathbf{x},\mathbf{y})\mathbf{u}(\mathbf{y},s)\right)d\Gamma = \int_\Gamma \mathbf{U}(\mathbf{x},\mathbf{y},s)\mathbf{t}(\mathbf{x},s)d\Gamma, \mathbf{x},\mathbf{y} \in \Gamma, \Gamma = \partial\Omega, \quad (2)$$

where $\mathbf{U}(\mathbf{x},\mathbf{y},s)$ and $\mathbf{T}(\mathbf{x},\mathbf{y},s)$ are matrices of fundamental and singular solutions, respectively, $\mathbf{T}^0(\mathbf{x},\mathbf{y})$ contains isolated singularities, $\mathbf{x}$ is integration point, $\mathbf{y}$ is observation point, $\mathbf{u}$ is generalized displacement vector, $\mathbf{t}$ is generalized force vector.

To solve equation (2), the boundary surface $\Gamma$ is divided into the generalized eight-node quadrangular elements. Generalized boundary functions of the first kind are approximated bilinearly, and generalized boundary functions of the second kind are assumed to be constant over the element.

The discrete representation of BIE (2) is constructed at the interpolation nodes of unknown boundary functions (collocation points) and has the following matrix form

$$[\Delta\mathbf{G}]\{\mathbf{T}\} = [\Delta\mathbf{F}]\{\mathbf{U}\}. \quad (3)$$

Matrices $\Delta\mathbf{G}$ and $\Delta\mathbf{F}$ contain integrals of the components of matrices $\mathbf{U}(\mathbf{x},\mathbf{y},s)$ and $\mathbf{T}(\mathbf{x},\mathbf{y},s)$, multiplied by the shape functions. The choice of the numerical integration scheme for computing the integral depends on its type. When a collocation point lies on integration element, the procedure for revealing the feature is performed. To improve the accuracy of integration on an element that does not contain a collocation point, in addition to the Gauss integration formulas, a hierarchical integration algorithm is applied, wherein the element is subdivided until the specified accuracy is achieved.

The solution in the time domain is obtained using the time-step method of numerical inversion of the Laplace transform. This method is close in its formulation to the CQM, but, in contrast to it, is based on the operational calculus of integrating original $f(s)$ of representation $\hat{f}(s)$. In general, the integral

$$y(t) = \int_0^t f(\tau)d\tau \quad (4)$$

is approximated as follows [11]:

$$y(0)=0, y(n\Delta t)=\mathbf{b}^T\mathbf{A}^{-1}\sum_{k=1}^{n}\boldsymbol{\omega}_k^{\Delta t}, n=1,...N \tag{5}$$

where $N$ is number of equal time steps. In this expression the quadrature weights $\boldsymbol{\omega}_k^{\Delta t}$ are determined using Laplace representation $\hat{f}(s)$ and the Runge-Kutta method. The quadrature weights can be expressed by the Cauchy integral form and approximated by using a trapezoidal rule with the number of steps $L$ as follows:

$$\boldsymbol{\omega}_n^{\Delta t}=\frac{R^{-n}}{L}\sum_{l=0}^{L-1}\hat{f}\left(\frac{\psi(z)}{\Delta t}\right)\frac{\psi(z)}{\Delta t}e^{-nl\frac{2\pi}{L}\mathrm{i}},\ z=Re^{l\frac{2\pi}{L}\mathrm{i}}, \tag{6}$$

$$\psi(z)=\mathbf{A}^{-1}-z\mathbf{A}^{-1}\mathbf{1}\mathbf{b}^T\mathbf{A}^{-1}, \tag{7}$$

where $\psi(z)$ is characteristic function of the Runge-Kutta method and $\mathbf{1}=(1,1,...,1)^T$. The parameter $R$ can be calculated by

$$R^L=\sqrt{\varepsilon}, \tag{8}$$

where $\varepsilon$ is the error of the numerical calculation of equation (6).

The approximation used in deriving formulas (6), (7) is based on the $m$-stage Runge-Kutta method written down employing Butcher's table:

$$\frac{\mathbf{c}\,|\,\mathbf{A}}{\mathbf{b}^T},\ \mathbf{A}\in\mathbb{R}^{m\times m},\ \mathbf{b},\mathbf{c}\in\mathbb{R}^m. \tag{9}$$

A correct formulation of a time-step scheme requires that the method be A-stable and L-stable. In the assumption of $\mathbf{b}^T\mathbf{A}^{-1}=(0,...,0,1)$, the method is automatically L-stable. It is also important to note that the quadrature weights $\boldsymbol{\omega}_k^{\Delta t}$ and the characteristic function $\psi(z)$ are $m$-order matrices.

In the present study, Lobatto (Lobatto IIIC) scheme was chosen as a particular example of the Runge-Kutta schemes meeting the formulated conditions.

## 3. Results and discussion

The problem of the impact of force $F(t)=1N/m^2$ on compound prismatic poroelastic column with a rigidly fixed end is considered. The problem setting is shown in figure 1. Boundary-element meshes for each of the parts $\Omega_1$ and $\Omega_2$ contain 512 elements. Plots of displacement and pore pressure at the midpoint of the body are shown in figures°2–3. Comparison of the results, obtained using nodes of 2- and 3-stage Lobatto methods, was carried out.

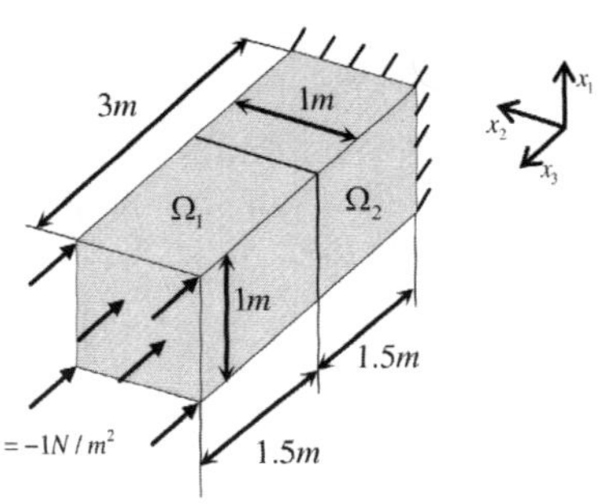


**Figure 1.** Geometry and boundary conditions of a partially saturated poroelastic column.

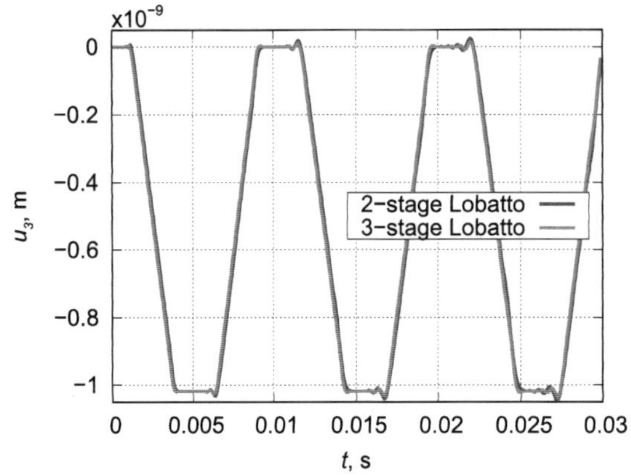


**Figure 2.** Displacement $u_3$ versus time at the midpoint of column.

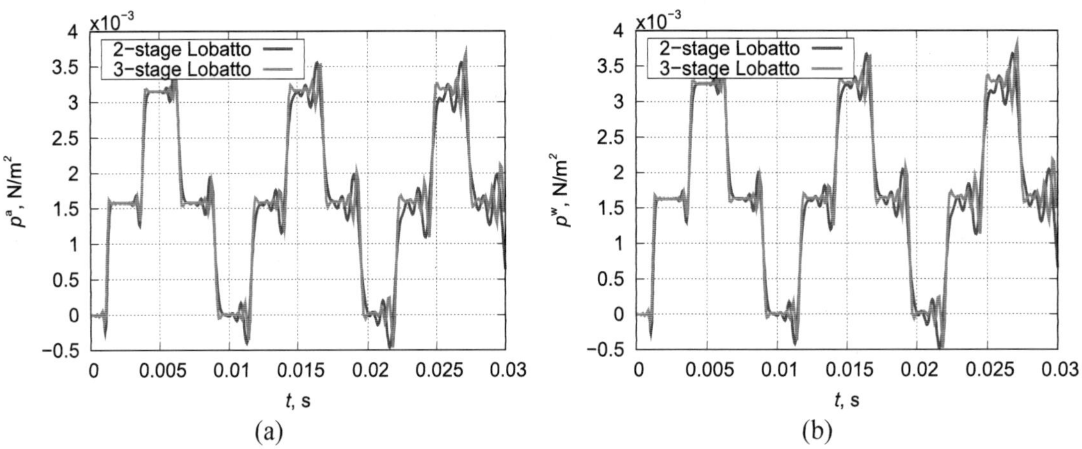


**Figure 3.** Pore pressures $p^a$ (a) and $p^w$ (b) versus time at the midpoint of the column.

Calculations using the 2-stage Lobatto method were performed with the time step value $\Delta t = 0.00003s$, and using the 3-stage Lobatto method with $\Delta t = 0.00012s$. Despite the significant difference in the time step values, the differences between the dynamic responses of the displacements obtained for each case are almost not noticeable on the plots. Fluctuating change of pore pressures allows comparing of the results in conditions of additional difficulties of approximation. The solution shown in figure 3, which obtained using three-stage method, is sufficiently smooth and retains the stable behavior on the considered time interval. In turn, the solution obtained using two-stage method, although smoother, but shows significant oscillations at the wave front points.

## 4. Conclusion

A boundary-element approach for solving the initial-boundary-value problems of the three-dimensional poroelastic bodies dynamics, which based on the application of the boundary elements method combined with the time-step scheme built on the Lobatto method nodes, is presented. Numerical studies of a model problem of the Heaviside function impact on a three-dimensional prismatic composite poroelastic body are carried out. A good agreement between the numerical boundary-element solution and the analytical solution is demonstrated.

**Acknowledgments**
The research was carried out under the financial support of the Russian Scientific Foundation (project no. 15-19-10056)

**Appendix A**

**Table 1.** Parameters of Massilon sandstone

| Parameter type | Symbol | Value | Unit |
|---|---|---|---|
| Porosity | $\varphi$ | 0.23 | - |
| Density of the solid skeleton | $\rho_s$ | 2650 | kg/m$^3$ |
| Density of the water | $\rho_w$ | 997 | kg/m$^3$ |
| Density of the air | $\rho_a$ | 1.10 | kg/m$^3$ |
| Drained bulk modulus of the mixture | $K$ | $1.02\times10^{9}$ | N/m$^2$ |
| Shear modulus of the mixture | $G$ | $1.44\times10^{9}$ | N/m$^2$ |
| Bulk modulus of the solid grains | $K_s$ | $3.5\times10^{10}$ | N/m$^2$ |
| Bulk modulus of the water | $K_w$ | $2.25\times10^{9}$ | N/m$^2$ |
| Bulk modulus of the air | $K_a$ | $1.10\times10^{5}$ | N/m$^2$ |
| Intrinsic permeability | $k$ | $2.5\times10^{-12}$ | m$^2$ |
| Viscosity of the water | $\eta_w$ | $1.0\times10^{-3}$ | Ns/m$^2$ |
| Viscosity of the air | $\eta_a$ | $1.8\times10^{-5}$ | Ns/m$^2$ |
| Gas entry pressure | $p^d$ | $5.0\times10^{4}$ | Ns/m$^2$ |
| Saturation degree | $S_w$ | 0.9 | - |
| Residual water saturation | $S_{rw}$ | 0 | - |
| Residual air saturation | $S_{ra}$ | 1 | - |
| Pore size distribution index | $\vartheta$ | 1.5 | - |

$$\kappa_a = K_{ra}k/\eta_a,\ \kappa_w = K_{rw}k/\eta_w,\ K_{rw} = S_e^{(2+3\vartheta)/\vartheta},\ K_{ra} = (1-S_e)^2\left(1-S_e^{(2+\vartheta)/\vartheta}\right) \quad (1)$$

$$S_e = \frac{S_w - S_{rw}}{S_{ra} - S_{rw}},\ S_u = -\frac{\vartheta(S_{ra}-S_{rw})}{p^d}\left(\frac{S_w - S_{rw}}{S_{ra} - S_{rw}}\right)^{\frac{\vartheta+1}{\vartheta}},\ \beta = \frac{\kappa_w \varphi \rho_w s}{\varphi S_w + \kappa_w \rho_w s},\ \gamma = \frac{\kappa_a \varphi \rho_a s}{\varphi S_a + \kappa_a \rho_a s}, \quad (2)$$

$$B_1 = G\nabla^2 - (\rho - \beta S_w \rho_w - \gamma S_a \rho_a)s^2,\ B_2 = K + \frac{G}{3},\ B_3 = -(\alpha-\beta)S_w,\ B_4 = -(\alpha-\gamma)S_a, \quad (3)$$

$$B_5 = -(\alpha-\beta)S_w s,\ B_6 = -\left(\zeta S_{ww} S_w + \frac{\varphi}{K_w}S_w - S_u\right)s + \frac{\beta S_w}{\rho_w s}\nabla^2,\ B_7 = -\left(\zeta S_{aa} S_w + S_u\right)s, \quad (4)$$

$$B_8 = -\left(\alpha-\gamma\right)S_a s,\ B_9 = -\left(\zeta S_{ww} S_a + S_u\right)s,\ B_{10} = -\left(\zeta S_{aa} S_a + \frac{\varphi}{K_a}S_a - S_u\right)s + \frac{\gamma S_a}{\rho_a s}\nabla^2, \quad (5)$$

$$\alpha = 1 - K / K_s,\ \zeta = (\alpha - \varphi) / K_s,\ S_{ww} = S_w - \vartheta(S_w - S_{rw}),\ S_{aa} = S_a + \vartheta(S_w - S_{rw}). \quad (6)$$

## References

[1] Igumnov L, Ipatov A, Belov A and Petrov A 2015 *EPJ Web Conf.* **94** 04026
[2] Igumnov L A, Litvinchuk S Yu and Petrov A N 2016 *Mater. Phys. Mech.* **28** 96–100
[3] Vorobtsov I, Belov A and Petrov A 2018 *EPJ Web Conf.* **183** 01042
[4] Petrov A I and Belov A A 2018 *MATEC Web Conf.* **226** 03003
[5] Lubich C 1988 *Numer. Math.* **52** 129–145
[6] Lubich C 1988 *Numer. Math.* **52** 413–425
[7] Schanz M and Antes H 1997 *Meccanica* **32** 179–186
[8] Abreu A I, Carrer J A M and Mansur W J 2003 *Eng. Anal. Bound. Elem.* **27** 101–105
[9] Zhang Ch 2000 *Int. J. Solids Struct.* **37** 6107–6130
[10] Lubich C and Ostermann A 1993 *Math. Comp.* **60** 105–131
[11] Banjai L, Messner M and Schanz M 2012 *Comput. Methods Appl. Mech. Engrg.* **245–246** 90–101
[12] Calvo M P, Cuesta E and Palencia C 2007 *Numer. Math.* **107** 589–614
[13] Banjai L and Lubich C 2011 *BIT* **51** 483–496
[14] Banjai L, Lubich C and Melenk J M 2011 *Numer. Math.* **119** 1–20
[15] Banjai L and Sauter S 2008 *SIAM J. Numer. Anal.* **47(1)** 227–249
[16] Li P and Schanz M 2013 *Eng. Anal. Bound. Elem.* **37** 1483
[17] Igumnov L A, Litvinchuk S Yu, Tarlakovsky D V and Lokteva N A 2013 *Problems of Strength and Plasticity* **75(2)** 130–136

IOP Conf. Series: Materials Science and Engineering **510** (2019) 012029 doi:10.1088/1757-899X/510/1/012029

# Numerical model of optical fiber piezoelectric feedback detector used for aviation composite constructions elements' geometry control

**A A Pan'kov and P V Pisarev**

Perm National Research Polytechnic University, 614990, Komsomolskii ave., 29, Russia, Perm

E-mail: pisarev85@live.ru

**Abstract**. The authors developed a numerical 3D model of a piezoelectroluminescent fiber-optic sensor for diagnosing a volumetric stress-strain state in a helicopter blade with controlled geometry. The authors of the article have performed a numerical calculation of non-uniform coupled electro-elastic fields in the elements of a sensor fragment embedded in a fiberglass sheathing of the main helicopter propeller. The calculation was carried out taking into account the action of the control voltage on the sensor electrodes. The numerical values of the informative and control coefficients of the sensor necessary for diagnosing the components of the strain tensors at the macro and micro levels of the textile composite are calculated.

## 1. Introduction

High-loaded elements of aircraft structures, such as an airplane wing or helicopter blades, create significant oscillations and noise due to changes in the aerodynamic loads acting on them when their azimuthal angle changes. With the advent of piezo-active materials, it became possible to develop efficient intelligent systems to tackle vibrations. Piezoactuators, integrated into the wing skin or rotor blades, create dynamic twisting and curvature of the structure, adapting it at any time to the flight conditions, which significantly reduces vibrations, noise and, in general, improves the flight characteristics of the structure.

Modern high-precision sensors based on intelligent materials are needed to monitor structural deformation fields and provide feedback in the controlled elements of composite structures. Such sensors can be integrated into complex composite structures as sensitive, translational and control elements that adapt to given conditions. For these purposes, fiber optic sensors are widely used, in which fiber is the main element. Optical fiber can be used as a transmission line of an informative signal and / or a sensitive element, in particular, in the form of a section of an optical fiber with a Bragg diffraction grating [1-4]. A promising solution to the problem of feedback in the elements of composite structures are piezoelectric sensor networks [5-8], embedded in the structure. New designs of piezoelectroluminescent fiber-optic sensors for diagnosing pressure [9], complex volumetric stress-strain state [10] and temperature [11] inside the composite structure were proposed in the patents [9-11]. These designs implement the algorithms [12, 13] of the receiver-analyzer processing integrated optical signals at the output of the sensor optical fiber.

The goal of this work is to develop a numerical model in ANSYS and to study the patterns of distribution of bounded electroelastic fields in the elements of a piezoelectric-luminous optical fibersensor [10] embedded inside a representative element of a deformable composite structure with its complex volumetric stress-strain state.

## 2. Model of the sensor

Piezoelectroluminescent fiber-optic sensors [9-12] are intended for the specified diagnosis of pressure distribution [6, 12] and/or temperature [11] and complex volume stress-strain state $\boldsymbol{\sigma}^*$ inside composite structures (Figure 1) [8] based on the results of processing a receiver-analyzer of the intensities $I$ of the integrated light signals at the output from the sensor fiber. The sensor (Figure 1) is an optical fiber 1 with coaxial electroluminescent 2 and piezoelectric 3 layers.

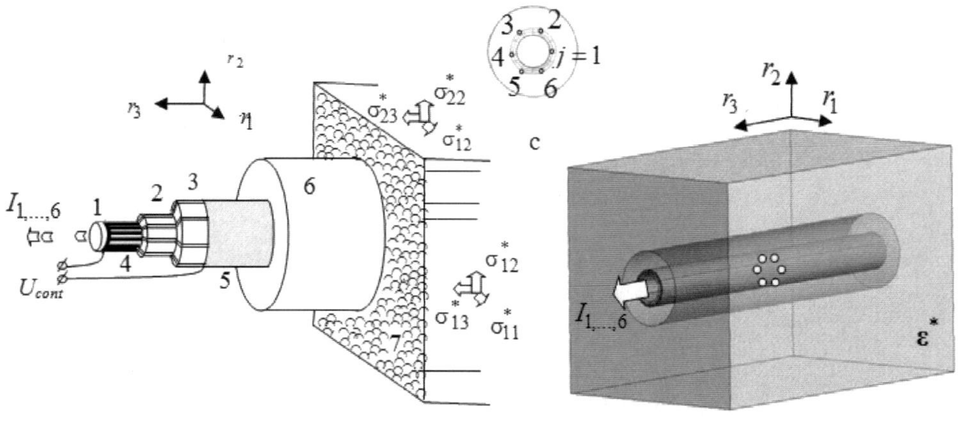


**Figure 1.** Piezoelectroluminescent fiber-optic sensor of the volumetric stress state (a), computational domain with a sensor fragment (b), test points in the cross section of the sensor (c)

Optical fiber and electroluminescent layer are separated by an internal translucent or perforated control electrode 4, an external control electrode 5 is located on the surface of the piezoelectric layer. The buffer layer 6 is required for mechanical translation of the sensor piezoelectric elements of a homogeneous macroscopic (averaged) component of the micro-inhomogeneous stress-strain state from the loaded composite region 7 in order to eliminate the "parasitic" effect of random pulsation fields of the microstructure in the region 7 of the sensor vicinity to the measurement results. In the sensor (Figure 1), the electroluminescent and piezoelectric layers are divided into geometrically equal radial-longitudinal boundaries shared by both layers. In particular, for six-parameter sensors, six "measuring elements" are distinguished - cylindrical two-layer sectors. In the sensor's measuring elements directions of the piezoelectric phases (piezoelectric elements) spatial polarizations and the frequency of the light output electroluminescent phases are different in all six sectors. The polarization directions of the piezoelectric elements are determined from the condition of non-coplanarity of the polarization directions for arbitrary three sensor sectors. Piezoelectric elements can be either different or same piezoelectric, in particular, PVDF polymer transversely isotropic material, but with different spatial polarization directions by sector. The direction of polarization is the axis of symmetry of the PVDF piezoelectric. Informative light signals occur in the local areas of the measuring elements of the sensor due to the mechanoluminescent effect (due to the interaction of the piezoelectric 3 and electroluminescent 2 sensor elements), penetrate the optical fiber 1 through the translucent electrode 4 and are transmitted via optical fiber 1 to the receiver-analyzer. The control voltage $U_{\text{cont}}$ on the electrodes 4, 5 makes it possible to identify the inhomogeneities locations of the diagnosed characteristics, in particular, the deformation fields in composite structures, using scanning algorithms [12].

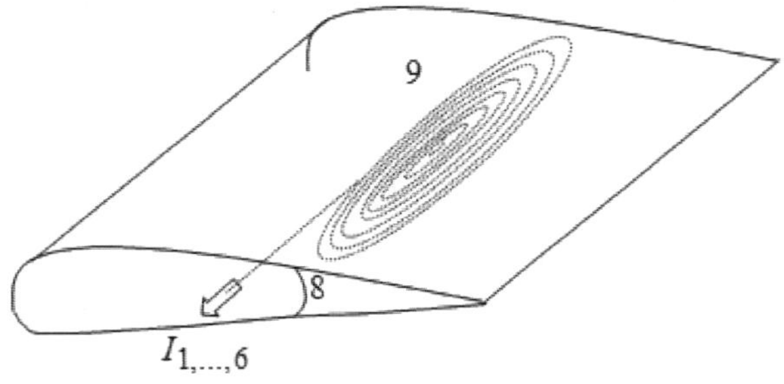


**Figure 2.** Fiber-optic sensor 9 inside the near-surface skin of the helicopter blade with a controlled piezo-actuator 8 geometry

The sensor (Figure 1) is characterized by its control$\mathbf{a}_1$ and informative$\mathbf{a}_2$ transfer coefficients, which establish a linear relationship between the electrical voltages on the electroluminescent layer

$$U_{lum(j)} = \alpha_{1(j)} U_{cont} + \alpha_{2(j)mn} \varepsilon^*_{mn} \quad (1)$$

(on each of the six elements of the electroluminescent layer $U_{lum(j)}$ in the $j$-th circular sector) with the amount of control voltage $U_{cont}$ applied to the electrodes and diagnosed parameters $\varepsilon^* = \{\varepsilon^*_{11}, \varepsilon^*_{22}, \varepsilon^*_{33}, \gamma^*_{23}, \gamma^*_{13}, \gamma^*_{12}\}$ in the cross section of the sensor with coordinate$r_3$, where the desired axial $\varepsilon^*_{11}, \varepsilon^*_{22}, \varepsilon^*_{33}$ and shear deformation $\gamma^*_{23}, \gamma^*_{13}, \gamma^*_{12}$ of the diagnosed complex volume deformed states are calculated in the considered location inside the composite structure, $j = \overline{1.6}$.

Figure 2 schematically depicts a helicopter blade [14] with a bimorph piezo-actuator of flexural type 8 located inside the rear part of the blade profile, supplemented with a fiber-optic sensor 9 with spiral stacking inside the near-surface (in particular, local) area of the skin with controlled geometry. Feedback is necessary to adapt the operating mode of the piezo-actuator (for example, the magnitude and location of the controlled force effect of the piezo-actuator on the actuator and the blade profile) to the special aspects (including random weather conditions) of the structure's service.

## 3. Numerical simulation

Geometrical parameters of the numerical model of the sensor (Figure 1): radii of concentric cylindrical surfaces $r_{(1)} = 1$ mm, $r_{(2)} = 1.2$ mm, $r_{(3)} = 1.4$ mm, $r_{(4)} = 2.8$ mm directed along the$r_3$axis in the center of the parallelepiped with edges oriented along the coordinate axes $r_i$ and with values of 16.8 mm in the $r_{1,2}$ axis and 27.2 mm in the $r_3$ axis; the computational domain was discretized into $18 \cdot 10^6$ finite elements, of which $12 \cdot 10^6$ stand for the cylindrical region of the sensor itself, taking into account the buffer layer. Electroelastic characteristics of the sensor phases: the isotropic properties of the fiber, the isotropic properties of the polymer electroluminescent phosphor, the transversely isotropic properties of the PVDF polymer piezoelectric (in the main axes) are given in [15], the isotropic elastic properties of the polyethylene buffer layer in [16], the transversely isotropic with a symmetry axis$r_3$ parallelepiped properties are equal to the effective properties of a unidirectional fiberglass plastic with a 0.6 volume fraction of fibers.

To calculate the informative coefficients $\mathbf{a}_{\varepsilon(j)}$, simple deformations $\varepsilon^*_{ij}$ of the parallelepiped were sequentially set through the displacements $\hat{u}_i = \varepsilon^*_{ij} \hat{r}_j$ of points $\hat{r}$ on its faces for given zero values of the control potential of the sensor electrodes. To calculate the control coefficients $a_{U(j)}$, we considered the case of the effect of a unit voltage $U_{cont}$ on the control electrodes in the absence of displacements for the points of the parallelepiped, $\boldsymbol{\varepsilon}^*=0$. According to the results of numerical simulations, numerical values of the informative $\mathbf{a}_{\varepsilon(j)}$ and $\mathbf{a}_{U(j)}$ control coefficients of the sensor (Figure 2) are needed to diagnose a complex volumetric stress state inside composite structures, the find locations and inhomogeneity

IOP Conf. Series: Materials Science and Engineering **510** (2019) 012029 doi:10.1088/1757-899X/510/1/012029

values of the components of stress and strain tensors at macro and micro levels of the composite diagnostic algorithms [12].

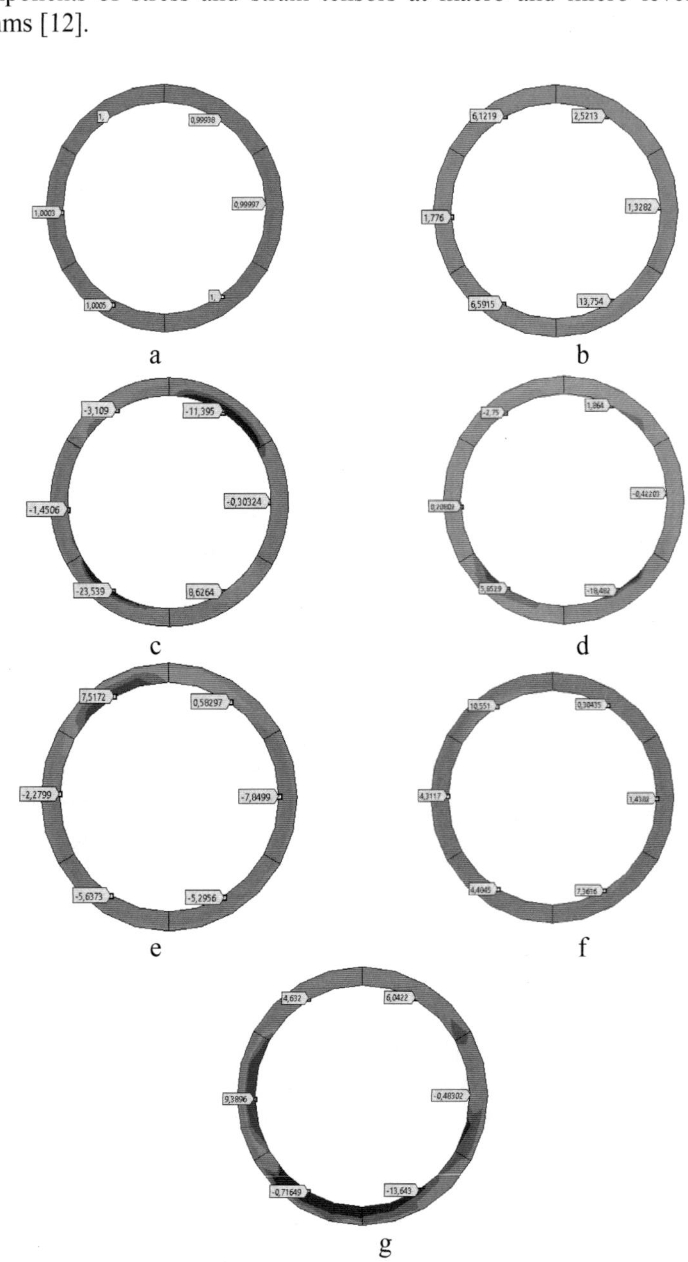


**Figure 3.** Electric potentials in the indicator sectors of the sensor for the given control voltage $\hat{U}_{cont} = -1$ V ($\boldsymbol{\varepsilon}^*=0$) (a), electric potentials for the axial deformations $\varepsilon^*_{11}$ (b), $\varepsilon^*_{22}$ (c), $\varepsilon^*_{33}$ (d) equalling to $10^{-5}$,electric potentials for the shear angles $\gamma^*_{23}$ (e), $\gamma^*_{13}$ (f) equalling to $0.6176 \cdot 10^{-5}$, $\gamma^*_{12} = -10^{-5}$(g) electric potentials in the indicator sectors of the sensor for zero control voltage ($\hat{U}_{cont} = 0$ V)

IOP Conf. Series: Materials Science and Engineering **510** (2019) 012029 doi:10.1088/1757-899X/510/1/012029

Thus, when solving the problem of diagnosing a stress-strain state in composite structures, the desired components of the macrodeformation tensor **ε*** of the composite are determined from the solution of a system of linear algebraic equations

$$[A]\begin{Bmatrix} \varepsilon_{11}^{*} \\ \varepsilon_{22}^{*} \\ \varepsilon_{33}^{*} \\ \gamma_{23}^{*} \\ \gamma_{13}^{*} \\ \gamma_{12}^{*} \end{Bmatrix} = \{\Delta\} \tag{2}$$

found using algorithms [12] (for processing the intensities of optical signals at the output of the sensor optical fiber with varying control voltage $\hat{U}_{cont}$) components $\Delta_{(j)}$, taking into account the control coefficients of the sensor $a_{U(j)}$ (1).

## 4. Conclusion

A numerical 3D model of a piezoelectroluminescent fiber optic sensor has been developed for diagnosing the volumetric stress-strain state in a helicopter blade with controlled geometry. A numerical calculation of non-uniform coupled electro-elastic fields in the sensor fragment elements embedded in the fiberglass sheathing of the helicopter rotor is performed, taking into account the effect of the control voltage on the sensor electrodes. The numerical values of the informative and control coefficients of the sensor necessary for diagnosing the components of the strain tensors at the macro and micro levels of the textile composite are calculated.

## Acknowledgments

The study was carried out at the Perm National Research Polytechnic University with the support of the Russian Science Foundation (project No. 18-19-00722).

## References

[1] Okosi T 1990 *Fiber-optical sensors* (L.: Energoatomizdat) p 256

[2] Guemes A, Fernandez–Lopez A and Soller B 2010 *Structural Health Monitoring–an International Journal* **9 3** 233–245

[3] Pan'kov A A 2018 *Mechanics of Composite Materials* **54 4** 513-522

[4] Balta J A, Bosia F, Michaud V, Dunkel G, Botsis J and Manson J A 2005 *Smart Materials & Structures* **14 4** 457–465

[5] Williamson C, Fixter L, Clarke A. 2006 *Sensor and sensor array for monitoring a structure.* (Patent US 20060254366 A1) Application No. US20060254366A1 dated 16.11.2006

[6] Beard Sh J, Qing X, Chan H L, Zhang C and Chang F–K 2008 *Method and apparatus for detecting a load change upon a structure and analyzing characteristics of resulting damage* (Patent US 7458266 B2)

[7] Sundaresan M J, Ghoshal A and Schulz M J 2002 *Sensor array system* (Patent US 6399939 B1)

[8] Shvetsov S N, Flek M B, Bragin S A and Akopjan V A 2009 *Izvestia of Samara Scientific Center of the Russian Academy of Sciences* **11 3(2)** 554-563

[9] Pankov A A 2017 *Fiber optic pressure sensor* (Patent RU №2630537)

[10] Pankov A A 2018 *Fiber-optical sensor of volumetric stress state* (Patent RU №2643692)

[11] Pankov A A 2018 *Temperature sensor* (Application No. 201122892 of 06/22/2018)

[12] Pan'kov A A 2017 *Mechanics of Composite Materials* **53 2** 229-242

[13] Shvetsov S N, Solov'ev A N and Bragin S A 2008 *Proc. of the 4th International Conference on Noise and Vibration Engineering* pp 360-321

[14] Gromann B, Mauher K, Altmicus A and Mangelsdorf W 2011 *Propeller blade for a rotorcraft* (Patent RU №2412866)
[15] Sessler G M 1981 *Journal Acoust. Soc. Amer.* **70** **6** 1596-08
[16] Turik A V and Radchenko G S 2003 *Physics of a Solid* **45** **9** 1667-79

# Numerical calculation of technological parameters for automated layup of a thermoplastic composite construction

**A N Anoshkin, P V Pisarev and Yu S Kirova**

Perm National Research Polytechnic University, 614990, Komsomolskii ave., 29, Russia, Perm

E-mail: pisarev85@live.ru

**Abstract**. The authors of this work present the results of a research aimed to help choosing the technological parameters, such as speed, power and surface area of the heating source, for the automated layup process. These characteristics ensure bonding of thermoplastic composites layers to each other. As a result of the research, a mathematical model was developed for heating a tape of a thermoplastic composite material with a laser source. In the framework of numerical experiments, two computational cases were considered for PEEK (polyetheretherketone) and PPS (polyphenylene sulfide) materials. According to the results of numerical simulations, temperature distribution fields over the tape of thermoplastic composite materials were obtained. The technological parameters at which the process of bonding the tapes of a thermoplastic prepreg along the adjoining faces of the surface occurs were determined.

## 1. Introduction

In recent years, the scope of composite materials application has increased significantly. First of all, due to their wide use in civil aviation and engine building. The use of composites allows significantly reducing weight (for a number of parts in several times), improving environmental parameters and efficiency of the aircraft [1, 2]. In particular, there is an increased interest in the use of composite materials with thermoplastic matrices (TCM) in the production of a wide range of products, including critical parts of aviation equipment [3 - 5]. However, the automated layup of thermoplastic prepreg has a number of problems associated with the choice of process parameters, such as the tension force of the tape, the temperature in the area of contact of the tape with the tooling, the force of pressing the tape to the tooling, the choice of the tooling material. For this work, a research was conducted on the choice of the heating mode for the thermoplastic prepreg tape of two butt layers placed on a metal tooling. To determine the required speed of the laser source, we numerically simulated the process of heating the contacting layers of the thermoplastic prepreg to the molding temperature.

## 2. Numerical simulation

The object of the research is a tape of a thermoplastic composite material made of unidirectional TCM [6]. Four options for prepreg placement were simulated. For options 1 and 2, the general scheme of the geometric model is presented in Figure 1a, and for options 3 and 4 in Figure 1b. The constructed geometric models differed in the area of impact of the surface heat source. Geometrical dimensions of the tape: 7.5x900x0.15 mm. In the process of automated layup, the tape is laid over the first (support) layer of a thermoplastic located on a rectangular tooling, then next to it a second tape of a thermoplastic is laid on the support layer placed nearby. A laser beam heats the area of contact between the upper and lower layers of prepreg tapes. For options 1 and 2, the size of the contact spot was: width 7.5 mm, length

9.375 mm. For options 3 and 4, the dimensions of the contact spot were: width 7.5 mm with the addition of 0.7 mm and on each side, the length was 9.375 mm. Simulation of the motion of a laser contact spot is carried out in a fixed coordinate system. For the simulation, several thermoplastic materials PEEK and PPS stacked on metal tooling with geometrical dimensions 360x1500x5 mm were considered. In total, four calculation options were considered. The speed of the laser source was 1 m/s. After calculations, laser transit time for each section was obtained for simulation, the transit time for each section is presented in Figure 2.

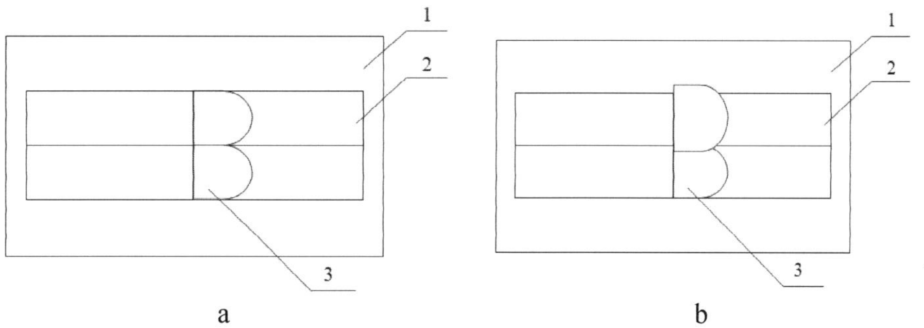


**Figure 1.**Scheme for simulation models: 1 - tooling, 2 - prepreg tape, 3 - laser focal spot.

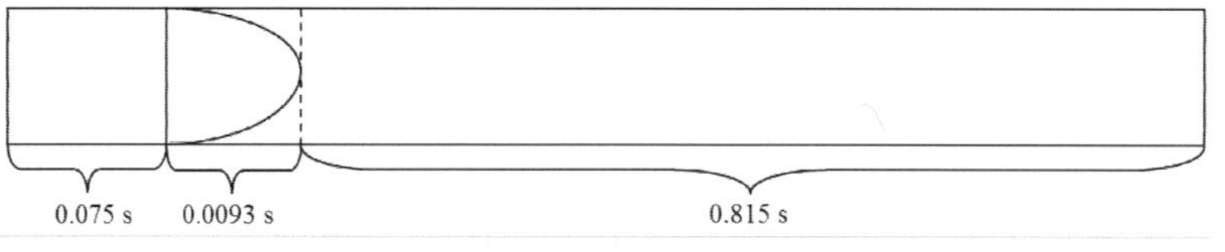


**Figure 2**. The time step of the simulation with automated display.

In this research the problem of numerical calculation of the temperature fields $T(x,y,z,t)$ in the prepreg tapes and tooling is posed. The processes are considered in a three-dimensional, non-stationary formulation, since the intensity of heat transfer between the layers of the prepreg and the tooling depends on the heating depth of the prepreg. A geometric model was developed and constructed to carry out numerical simulation of the heating of laminated CM with a thermoplastic matrix in the process of automated manufacturing (Figure 3).

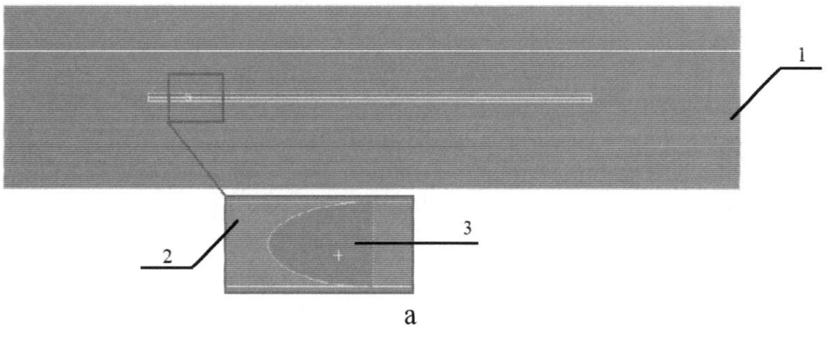

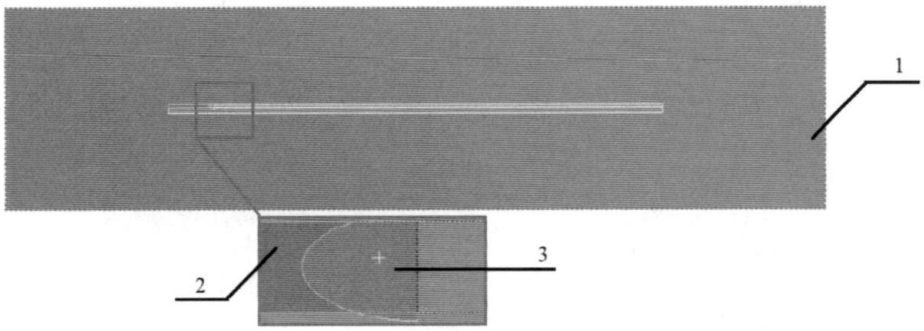

b

**Figure 3.** General view of the first (a), second (b) geometrical model: 1 - tooling; 2 – belt prepreg; 3 - laser focal spot.

The mathematical formulation of the problem includes the nonstationary heat conduction equation for anisotropic bodies [7].

$$C_p\rho\frac{\partial T}{\partial t}=\lambda_{11}\left(\frac{\partial T}{\partial x}\right)^2+\lambda_{22}\left(\frac{\partial T}{\partial y}\right)^2+\lambda_{33}\left(\frac{\partial T}{\partial z}\right)^2+(\lambda_{23}+\lambda_{32})\frac{\partial T}{\partial y}\frac{\partial T}{\partial z}+(\lambda_{31}+\lambda_{13})\frac{\partial T}{\partial z}\frac{\partial T}{\partial x}+ \\ +(\lambda_{12}+\lambda_{21})\frac{\partial T}{\partial x}\frac{\partial T}{\partial y} \quad (1)$$

where $T(x)$ is the value of the temperature field; $\lambda_{11}$, $\lambda_{22}$, $\lambda_{33}$, $\lambda_{23}$, $\lambda_{32}$, $\lambda_{31}$, $\lambda_{13}$, $\lambda_{12}$, $\lambda_{21}$are the thermal conductivity of the material in the *x,y,z* directions; *t* is for time; $C_p$ is the specific heat capacity of the material; ρ is the density of the material.

As initial conditions, the value of the temperature of the thermoplastic, tooling and external environment was set $T(x,y,z,0)=22°C$. At the place of contact of the layers and in the interaction area of the lower layer and the tooling, boundary conditions of the fourth kind were assigned (2). Between the lower edge of the upper layer and the upper edge of the lower layer of thermoplastic there is a contact spot with a constant heating temperature $T(x,y,z,t)=420°C$ for the PEEK material and $T(x,y,z,t)=280°C$ for the PPS material. In the area of contact of the prepreg layers and the tooling with the external environment, the boundary conditions of the third kind were assigned (3).

$$\lambda_1\frac{\partial T_1(t)}{\partial n}\overline{l}_n=\lambda_2\frac{\partial T_2(t)}{\partial n}\overline{l}_n, \quad (2)$$

where $\lambda_1$, $\lambda_2$ are the thermal conductivity coefficients of the first and second body; $T_1$, $T_2$- the temperature of the first and second bodies on the border of their contact; $\overline{l_n}$ is the vector of normal to the surface of the body.

$$\lambda\frac{\partial T}{\partial n}=h(T_n(M,t)-T_{oc}), \quad (3)$$

where $T_n$ - the temperature of the thermoplastic layer; $T_{oc}$ –the value of the external environment temperature; *h* is the coefficient of heat transfer between the external environment and the surface of the body; *M* is the point belonging to the surface; $l_n$is the normal vector to the surface of the body.

While constructing and local grinding of the mesh, no sharp differences in the geometric dimensions of the neighboring elements were allowed (sharp differences are understood to be more than 2 times). The mesh was adapted in the areas of high temperature gradients, namely, on the contact surface of the laser spot and the prepreg tape. The maximum element size for prepreg tape was 1 mm, for tooling - 2 mm. (Figure 4). The total number of finite elements was 2000 thousand elements. The final elements of the tooling are hexagonal in shape. The elements of prepreg tape represent the ethereal form.

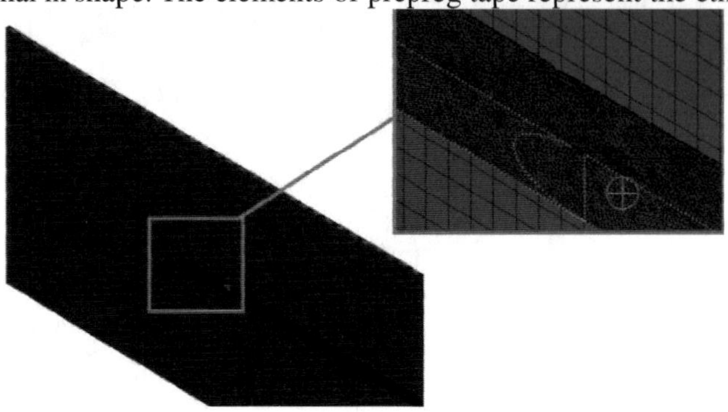

**Figure 4.** General view of the finite element model of a layered blank placed side by side on a flat metal tooling during automated fabrication

In the numerical experiments, the process of automated calculation of thermoplastic PEEK and PPS materials was simulated. During the processing, the prepreg tape was laid on a metal tooling. A moving laser beam heated the area of contact between the upper and lower layers of prepreg tapes. The thermo physical properties of anisotropic thermoplastic prepregs were determined analytically from the properties and relationships presented in [8, 9]. In the modeling process, it was assumed that one heat source should be considered as two layers of thermoplastic, neglecting the convection between the layers in the areas of laser spot impact, because the contact between the bodies is considered as perfect.

To implement the sequential layout of prepreg tapes, taking into account interlayer heat transfer, an algorithm and a software module integrated into the ANSYS engineering analysis system have been developed and implemented. The algorithm provides the transfer of data obtained as a result of the solution at the previous step for a new geometric model.

According to the results of numerical simulations, temperature distribution fields were obtained, for the convenience in determining the degree of the layers' heating, the temperature diagrams for the layers of thermoplastic tape were constructed. For the TCM with the PEEK matrix, the diagrams are shown in Figure 5. Figure 6 shows the temperature distribution diagrams for the TCM with the PPS matrix.

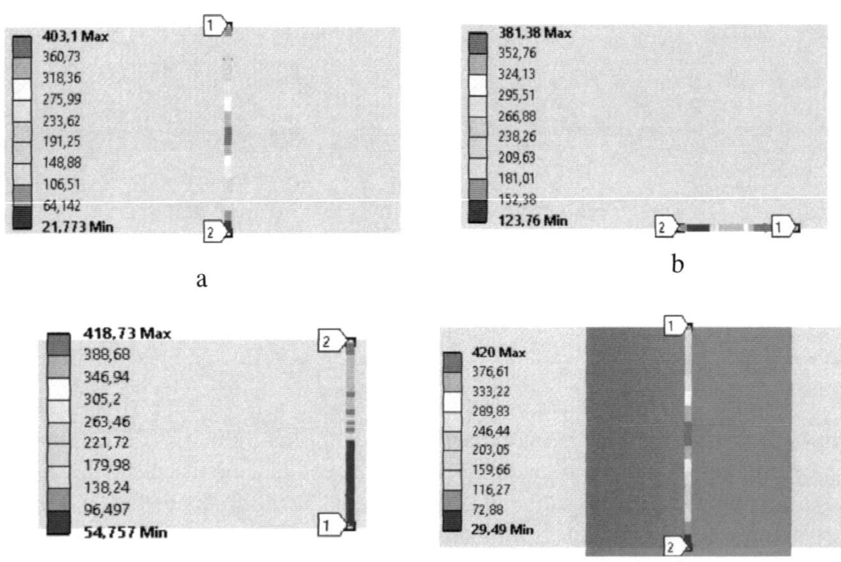

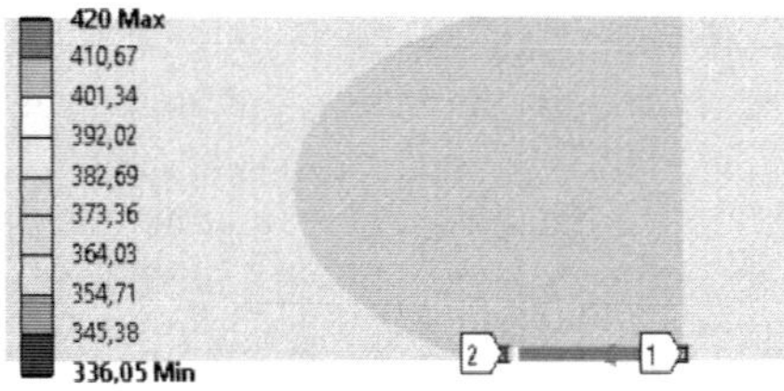


e

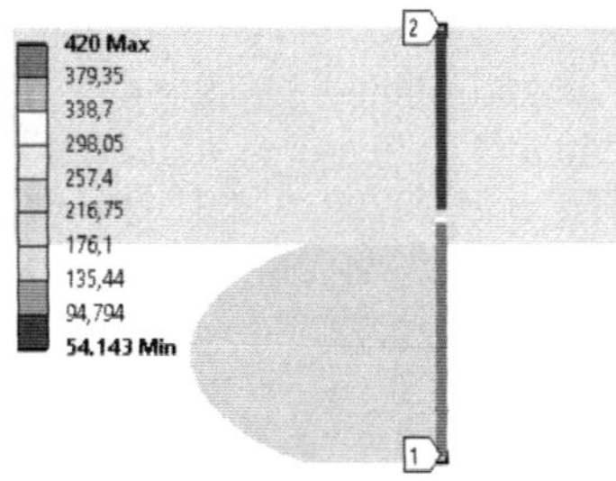


f

**Figure 5.** Temperature distribution epures and diagrams for temperature dependence on the height, length, and width for a PEEK material, g - along the height of two layers at a point close to the junction of the two layers placed side by side; b, d - on the border of the layers placed side by side on the surface of the first layer; b, e - the intersection of the width of two layers placed side by side on the surface of the first layer.

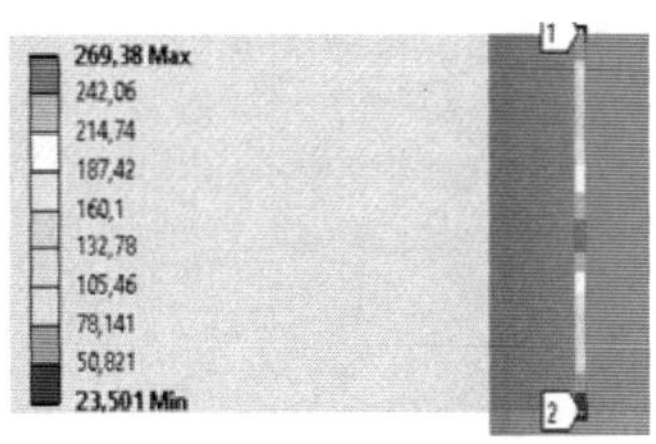


a

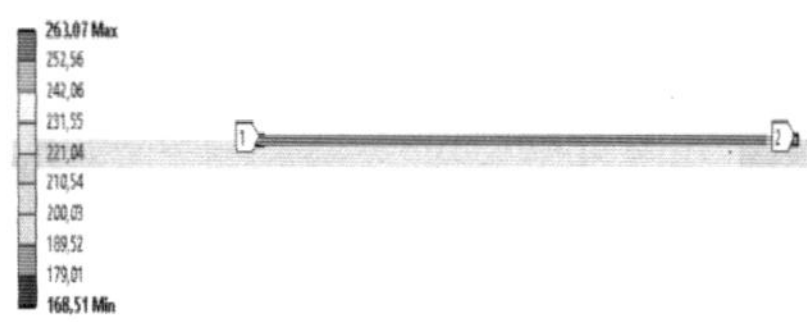

b

c

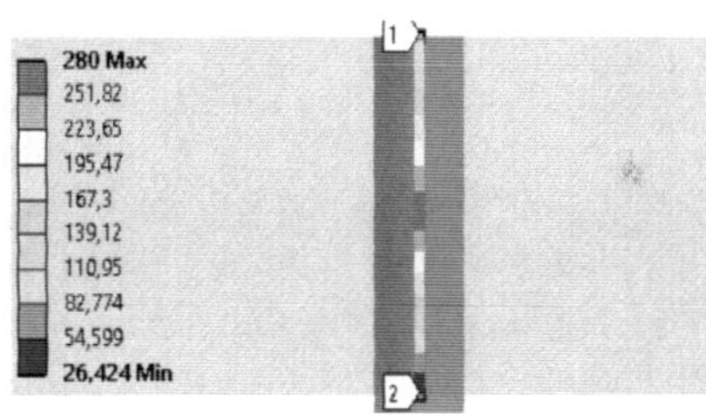


d

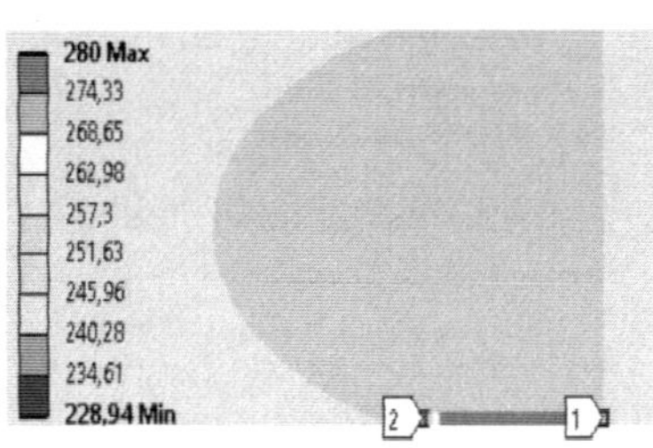


e

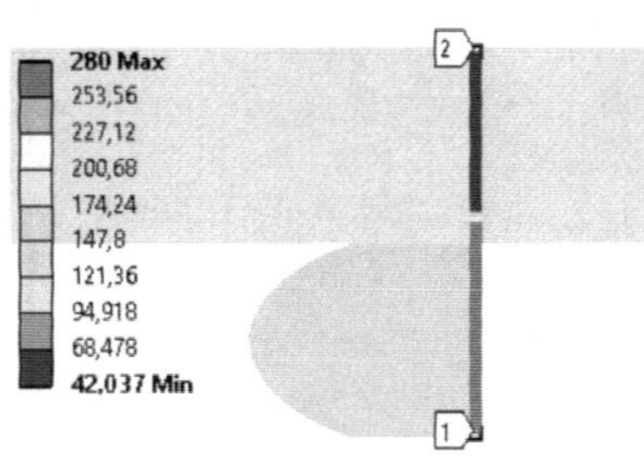


f

**Figure 6.** Temperature distribution epures and diagrams for temperature dependence on the height, length, and width for a PPS material, g - along the height of two layers at a point close to the junction of the two layers placed side by side; b, d - on the border of the layers placed side by side on the surface of the first layer; b, e - the intersection of the width of two layers placed side by side on the surface of the first layer.

The analysis of the plotted temperature distribution epures along the height of the TCM layer revealed that for model 1, heating to the molding temperature in the area of the contact spot reaches 12.5% for the PEEK material and 12.91% for the PPS material. For model 2, the values are 17.08% and 14.58% for the PEEK and PPS materials, respectively. Thus, we can conclude that increasing the width of the contact spot by 18.6% leads to the layers bonding together along the side faces, while simultaneously increasing the heating depth of the material. It should be noted that the use of thermal insulation coating is recommended to prevent overheating of the support layer. To ensure the recommended temperature processing and control overheating requires the use of thermal imaging control on the surface of the tape and tooling. According to the results of the conducted research, it was revealed that the speed of the laser source movement of 1 m/s is sufficient for heating the layers in the contact spot area for PEEK and PPS materials. This value was determined based on the speed of movement of the laser head of the equipment used for laying the composite tapes.

Thus, in the framework of the research, a mathematical model has been developed that describes the non-stationary processes of heating the tapes of a thermoplastic prepreg in the process of automated layup. The speed of the laser source was selected, to provide the temperature needed for molding of the considered materials. The conducted researched revealed, that a heating contact spot when having a large heating area leads to bonding of the first and second layers when arranging layers on each other and to bonding on the lower face of the second layer. Metal tooling has a greater heat capacity, it contributes to heating the thickness of the tooling, while carbon fiber tooling, in turn, removes heat better, thus preventing the material from heating. Interlayer bonding depends on the speed of the source and its power.

**Acknowledgments**
The work was carried out with the financial support of the state represented by the Ministry of Education and Science of Russia, the unique identifier of the project RFMEFI57717X0261

**References**

[1] Anoshkin A N, Pisarev P V and Kirova Y S 2018 *MATEC Web of Conferences* **224** 01080 (2018)
[2] Demkovich N A, Volkov I A and Yablochnikov E I 2016 *Proc. of the Samara Scientific Center of the Russian Academy of Sciences* **18 4(3)** 459-463
[3] Bader E Y, Petrova G N, Izotova T F and Gureeva E V 2013 *Trudy VIAM* **11** 1
[4] Kim H, Yarlagadda Sh, Gillespie J W, Shevchenko N B and Fink B K 2002 *Advanced Composite Materials* **11(1)** 71-80
[5] Mohannad A 2014 *Theses and Dissertations* p 652
[6] Gusev Yu A, Borshchev A V and Khrulkov A V 2015 *Trudy VIAM* **3** p 6
[7] Carslow G and Yeher D 1964 *Thermal conductivity of solids* (Nauka) p 231
[8] Fedulov B N, Saonov A A, Kantor M M and Lomov S V 2017 *Composites and nanostructures* **9(2)** 102-122
[9] Chukov D I, Stepashkin A A, Tcherdyntsev V V, Olifirov L K and Kaloshkin S D 2017 *Journal of Thermoplastic Composite Materials* 882-895

# Structural and luminescent characteristics of YAG phosphors synthesized in the radiation field

**D A Mussakhanov[1], A T Tulegenova[2], V M Lisitsyn [1], M G Golkovsky[3], L A Lisitsyna[4], Kh A Abdullin[2], M B Aitzhanov[2], Zh Karipbayev[5], A Kozlovsky[5] and Yu I. Michailov[6]**

[1]National Research Tomsk Polytechnic University, Tomsk, 634050, Russia
[2]al-Farabi Kazakh National University, Almaty, 050040, Kazakhstan
[3]Institute of Nuclear Physics of Siberian Branch Russian Academy of Sciences, Novosibirsk, 630090, Russia
[4]Tomsk State University of Architecture and Building, Tomsk, 634003, Russia
[5]L.N. Gumilyov Eurasian National University, Astana, 010008, Kazakhstan
[6]Institute of Solid State Chemistry and Mechanochemistry, Siberian Branch of the Russian Academy of Sciences, Novosibirsk, 630128, Russia

E-mail: tulegenova.aida@gmail.com

**Abstract**. YAG:Ce, YAGG:Ce ceramics were obtained by sintering the oxide powders in the radiation field. The results of investigations of the structure, composition and luminescence of ceramics are presented. The luminescence characteristics of powders exactly correspond to the phosphors luminescence used in practice obtained by solid-state synthesis methods. It has been established that in the used radiation exposure modes the main factor determining the efficiency of the synthesis is the ionization density.

## 1. Introduction

YAG:Ce phosphors are the most promising for use in light-emitting diodes (LEDs). The phosphors are crystalline multicomponent systems. Their synthesis is usually carried out using solid-phase reactions [1–5].

The melting points of the components are high. Therefore, the reproducibility of the results of the synthesis is low: the elemental composition of microcrystals differs significantly from that in the mixture. The elemental composition of the phosphor of different batches changes even during synthesis under the same conditions and the same initial composition of the mixture. Fluorescent properties are also changing. This is explained by the strong influence on the result of the synthesis of uncontrolled deviations of technological modes in the synthesis process. Therefore, we are constantly working to improve the technology of phosphor preparation, the develop of new technologies. A particular difficulty in the preparation of the phosphor is the first stage - the production of ceramics by sintering the initial oxide powders.

This paper presents the results of studies of the structural and luminescent characteristics of phosphors synthesized using a fundamentally new method - sintering in the radiation field

## 2. Samples preparation

The charge was composed of a mixture of the following powders: $Y_2O_3$ (22 – 36 weight %), $Al_2O_3$ (56–62 weight %), $Ce_2O_3$ (2.4 – 9.1 weight %) and $Gd_2O_3$ (0 – 12 weight %). The resulting mixture was poured into cavities with a depth of 5 mm of the massive copper crucible. A powerful electron flow directed from the vacuum to the medium with atmospheric pressure through a differential vacuum pumping system consisting of three stages was sent to the crucible. The electrons flow with the energy of 1.4 MeV and a power density of 22–25 KW per $cm^2$ from the ELV-6 accelerator was scanned along the crucible at a speed of 1 cm $s^{-1}$. The electron beam with a cross section at the surface of the crucible 1 $cm^2$ melted the charge for 1 s, which, after exposure, quickly hardened, forming a ceramic sample with a ratio of oxide components set by the charge composition. After cooling, the samples were removed from the crucible.

Figure 1 shows photographs of samples of synthesized YAG:Ce ceramics having a volume in the range of 0.8 to 1 $cm^3$. Samples YAG:Ce had a yellow color. The samples containing gadolinium, as a modifier (YAGG: Ce), had dark yellow color. Excitation with a radiation chip with $\lambda = 460$ nm leads to luminescence of YAG: Ce in the green, YAGG: Ce - in the yellow region of the spectrum.

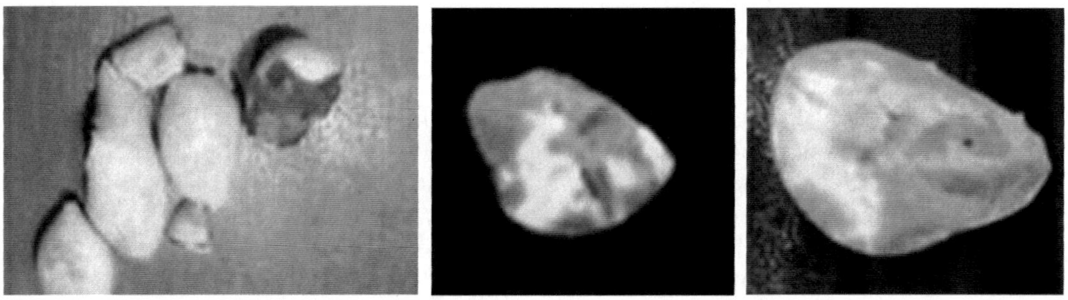

**Figure 1.** Photographs of samples of synthesized YAG: Ce ceramics illuminated by white light (the first picture) and excited by LED radiation (next two pictures)

From the pictures shown in the figure 1, it is clear that the luminescence is distributed non-uniformly over the sample surface. This is due to the inhomogeneous distribution of cerium oxide by volume of the charge and, accordingly, the dopant in terms of the volume of ceramics.

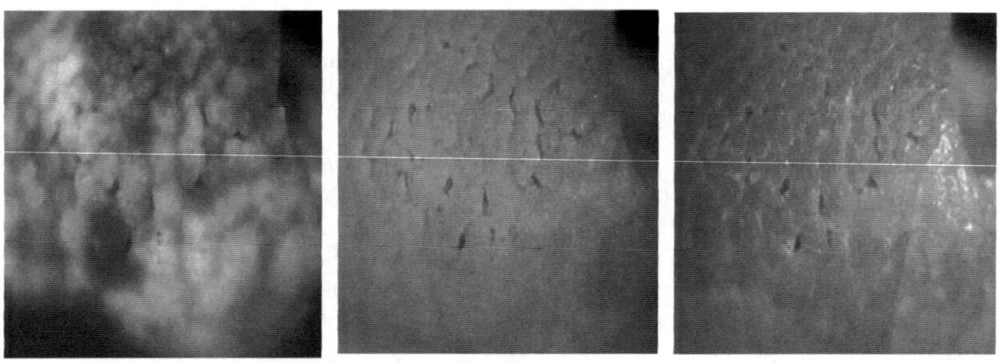

**Figure 2.** Photographs of one and the same surface area of a ceramic sample illuminated by light with $\lambda = 365$ and 450 nm from chips and an incandescent lamp (from left to right, respectively)

Figure 2 shows photographs of the one and the same surface area of ceramic sample illuminated by light from LED with $\lambda = 365$ and 450 nm and an incandescent lamp, respectively. The area of a picture is 3x2.5 mm$^2$. The synthesized sample of ceramics represents sticky fragments of a rounded shape. The fragments have a size of 100 - 200 microns. The fragments are firmly connected to each other. The photograph in white light shows that the crack passes along a plane unrelated to the shape of the fragment. Excitation by radiation from chips with $\lambda$ = 365 and 450 nm leads to the appearance of luminescence. The fields of the luminescent regions differ in shape and color. When excited by radiation from a chip with $\lambda$ = 365 nm, luminescence areas of yellow and purple are clearly distinguished, while excitation by radiation with $\lambda$ = 450 nm leads only to yellow luminescence almost uniformly distributed over the samples surface. The difference is probably due to the fact that irradiation with $\lambda$ = 365 nm excites, besides yellow, violet luminescence. The luminescence centers responsible for effective violet luminescence are unevenly distributed over the surface.

The surface state of the synthesized ceramics samples were studied using the Hitachi TM-3030 scanning electron microscope (SEM). The device made it possible to observe bulk samples with a shadow and volume contrast with a resolution of up to 30 nm. Figure 3 shows typical SEM images of the sample surface with an area of ~ 0.016 mm$^2$, magnified 1000 times. There are snapshots of the surface of a ceramics samples prepared from charges with different composition: of $Al_2O_3$ (56.8 %) + $Y_2O_3$ (34.1 %) + $Ce_2O_3$ (9.1 %) (3a) and $Al_2O_3$ (59.5 %) + $Y_2O_3$ (23.8 %) + $Gd_2O_3$ (11.9 %) + $Ce_2O_3$ (4.8 %) (3b).

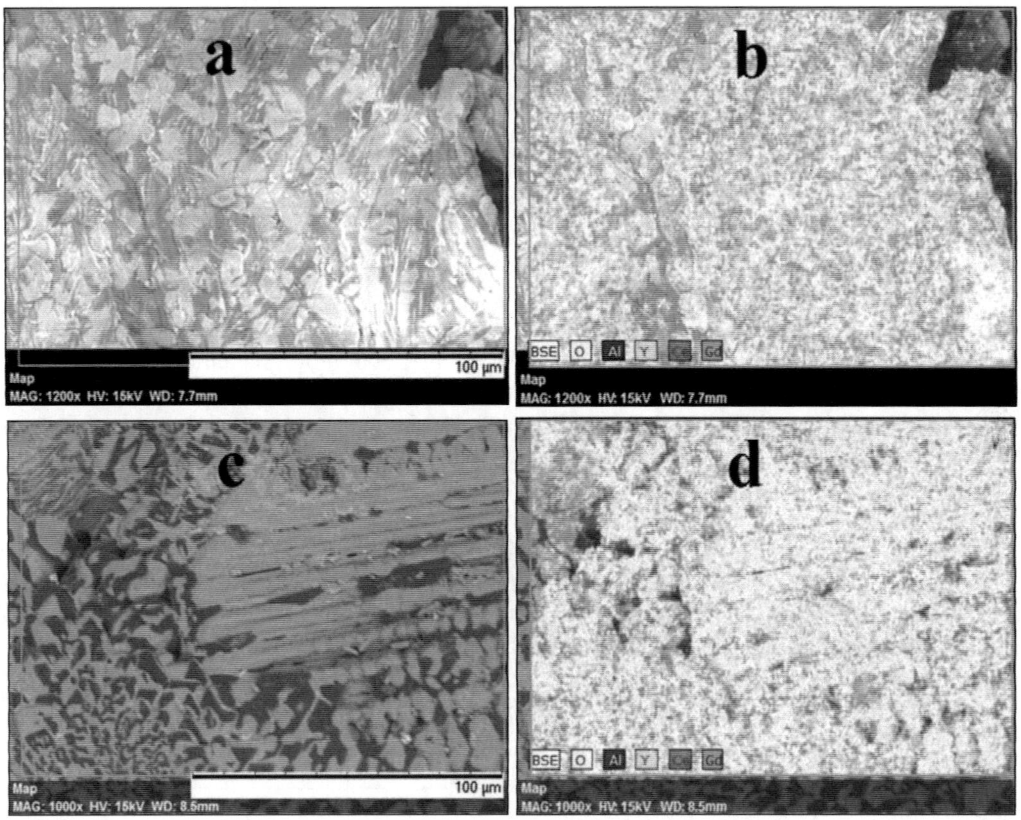


**Figure 3.** SEM images of samples surface with the different composition: (a) - $Al_2O_3$ (56.8 %) + $Y_2O_3$ (34.1%) + $Ce_2O_3$ (9.1 %); (c) - $Al_2O_3$ (59.5 %) + $Y_2O_3$ (23.8 %) + $Gd_2O_3$ (11.9 %) + $Ce_2O_3$ (4.8 %). Pictures b and d show the elemental composition distribution over the surface in these samples respectively

From presented images it is clear that the samples consist of glued particles, often of regular shapes. This suggests that crystal structure is formed under exposure to a radiation flow during the ceramics samples preparation.

## 3. Structure of synthesized YAG ceramics

The electron microscope with the Bruker XFlash MIN SVE has been chosen for the synthesized samples analysis. The elemental composition was determined in fragments of split samples of ceramics. The results of measurements are shown in Table 1. Samples 1 and 2 had the same initial composition of the charge. In table 1 different fragments of the same split sample of ceramics were represented. Ceramics sample 3 was prepared from the charge doped with Gd. The table 1 shows the elemental composition of different fragments of a crushed ceramic sample. The result presented in the table 1 show that the elemental composition in ceramic samples 1 and 2 is different, although the initial charge compositions, were the same. Compositional studies were conducted at different points on the plane of the fractured ceramic sample. There is a difference in the elemental composition over the surface. Consequently, during the synthesis of ceramics, a change in the elemental composition by volume occurs. There is a difference in the composition of the ceramic sample 3 from original composition of the charge.

**Table 1**. The elemental composition of the charge and ceramics samples prepared from it

| Sample | | Elemental composition, in % | | | | |
|---|---|---|---|---|---|---|
| | | O | Al | Y | Ce | Gd |
| 1 | In furnace-charge | 60 | 25 | 11.9 | 3.1 | - |
| | In ceramics | 57.4 | 33.3 | 7.55 | 1.8 | - |
| 2 | In furnace-charge | 60 | 25 | 11.9 | 3.1 | - |
| | In ceramics | 66.3 | 25.6 | 7.49 | 0.54 | - |
| 3 | In furnace-charge | 60 | 25 | 7.92 | 3.18 | 3.9 |
| | In ceramics | 52.8 | 37.7 | 7.29 | 1.35 | 0.8 |

Obviously, the heterogeneity of physical and chemical processes during of the synthesis takes place. It means that the charge could not be mixed.

Nevertheless, there is a correspondence between the composition of the elements in the charge and ceramics samples. Figures 3b and 3d show the measured distribution of the elemental composition on arbitrarily selected areas of the cleavage sample surface. As it can be seen from the presented images, the elements of the composition are distributed relatively homogeneously. This indicates that in the synthesis process in the radiation field, there is a fairly good mixing of elements.

X-ray diffraction analysis of ceramics samples was performed using a Rigaku Miniflex 600 X-ray diffractometer. It can be used for qualitative and quantitative phase analysis of polycrystalline materials as well.

The research results were presented in the form of ordinary spectrograms with high resolution. Table 2 shows the results of X-ray analysis in the form of a table with processed spectra. The table 2 also presents for comparison the results of X-ray diffraction analysis of phosphors of the SDL series and yttrium oxide powder. The space group of all phosphors and ceramics samples is Ia-3d and of yttrium oxide - Ia-3. The lattice parameters are represented in the table 1.

**Table 2**. X-ray analysis of prepared ceramics samples

| **Composition of sample** | **Type of phase** | **lattice parameters, Å** | **Note** |
|---|---|---|---|
| YAG:Ce<br>$Al_2O_3$ (59.5 %) + $Y_2O_3$ (23.8 %) +$Gd_2O_3$ (11.9%)+$Ce_2O_3$ (4.8 %) | $YGd_3$ $Al_5$ $O_{12}$:Ce | a= 12.064519<br>b= 12.064519<br>c= 12.064519 | 20 peaks corresponds to phases $YGd_3$ $Al_5$ $O_{12}$:Ce |
| YAG:Ce | $Y_3$ $Al_5$ $O_{12}$:Ce | a= 12.012482<br>b= 12.012482 | 20 peaks corresponds to phases $Y_3$ $Al_5$ $O_{12}$: Ce |

| | | | |
|---|---|---|---|
| $Al_2O_3$ (59.5 %) + $Y_2O_3$ (35.7 %) + $Ce_2O_3$ (4.8 %) | | c= 12.012482 | |
| $Y_2O_3$ – pure | $Y_2\ O_3$ | a= 10.615582<br>b= 10.615582<br>c= 10.615582 | 32 peaks corresponds to phases$Y_2\ O_3$ |
| SDL 2700 | $YGd_3\ A_{l5}\ O_{12}$:Ce | a= 12.074755<br>b= 12.074755<br>c= 12.074755 | 31 peaks corresponds to phases $YGd_3\ Al_5\ O_{12}$:Ce |
| SDL 3500 | $Y_3\ Al_5\ O_{12}$:Ce | a= 12.022043<br>b= 12.022043<br>c= 12.022043 | 20 peaks corresponds to phases $Y_3\ Al_5\ O_{12}$:Ce |
| SDL 4000 | $Y_3\ A_{l5}\ O_{12}$:Ce | a= 12.009555<br>b= 12.009555<br>c= 12.009555 | 36 peaks corresponds to phases $Y_3\ Al_5\ O_{12}$:Ce |

Thus, in the synthesized ceramics samples the main phase is YAG phase, just like in YAG phosphors.

## 4. Luminescence and excitation spectra of YAG ceramics

Luminescence spectra of samples and excitations spectra of the luminescence were measured with a CM 2203 spectrophotometer. Figure 4 shows the luminescence spectra of the synthesized ceramics samples excited by light with a wavelength of 345 nm. The luminescence spectra are characteristic of YAG:Ce and YAGG: Ce phosphors [6-8]. In the latter, some yttrium ions are replaced by gadolinium ions. It leads to shift of luminescence band to red region (figure 4). Luminescence spectra of ceramics samples of the same composition are similar.

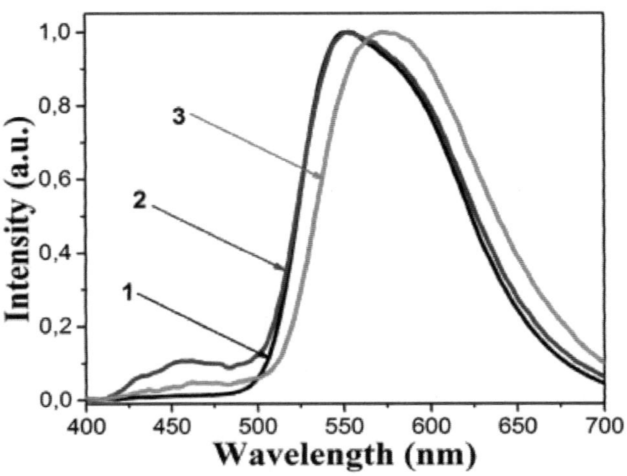


**Figure 4**. Luminescence spectra of ceramics samples of the different composition: (1, 2) - $Al_2O_3$ (59.5 %) + $Y_2O_3$ (35.7 %) + $Ce_2O_3$ (4.8 %); (3) - $Al_2O_3$ (62 %) + $Y_2O_3$ (25 %) + $Gd_2O_3$ (13 %) + $Ce_2O_3$ (1 %).

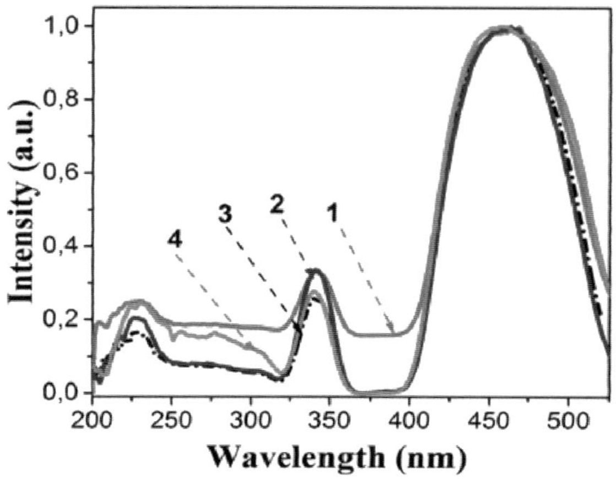


**Figure 5.** The excitation spectra of the emission in the ceramics samples of the different composition: (1) - $Al_2O_3$ (59.5 %) + $Y_2O_3$ (35.7 %) + $Ce_2O_3$ (4.8 %); (2) - $Al_2O_3$ (56.8 %) + $Y_2O_3$ (34.1 %) + $Ce_2O_3$ (9.1 %); (3) - $Al_2O_3$ (56.8 %) + $Y_2O_3$ (34.1 %) + $Ce_2O_3$ (9.1 %); (4) - $Al_2O_3$ (56.8 %) + $Y_2O_3$ (22.7 %) + $Gd_2O_3$ (11.4 %) + $Ce_2O_3$ (9.1 %).

Figure 5 shows excitation spectra of the luminescence at the maxima of the bands in the ceramics samples with different elemental composition. Three bands in the excitation spectra at 450, 340 and 220 nm in YAG:Ce phosphors are observed.

## 5. Discussion

The results of studies showed that ceramic based on yttrium-aluminum garnet, doped with cerium and synthesized in the radiation field, luminescence quite well when excited by radiation in a wide UV spectral range. The spectral characteristics of luminescence and excitation are quite similar to those measured for YAG:Ce phosphors. In the prepared ceramics the main phase is YAG: Ce (or YAGG: Ce) phase. Although the morphology of ceramics is very different from microcrystals of YAG: Ce phosphors, the luminescence characteristics indicate that nanodefects responsible for luminescence [9, 10] are formed in crystallites. Obviously, further standard processing of ceramics to transform it into a phosphor: crushing and annealing at temperatures of 1500–1700 K, will make it possible to obtain significantly better quantitative characteristics of the luminescence.

It is known that ionizing radiation fluxes are used in practice, for example, to modify material properties. In [11], it was shown that dielectric samples are destroyed when exposed to powerful short pulses of electron flow. In [12], the possibility of ceramics sintering under exposure to powerful fluxes of low-energy electrons, about 15 keV, is shown. Using a powerful flow of electrons with such low electron energies (penetration depth of 10 nm) that leads only to heating and emission. The mean of the free path of electronic excitations is about 10 nm as well. Therefore, to melt the refractory oxides, the flow of electrons can be directed not to the charge, but to the graphite crucible in which the charge is placed.

It is known to use powerful stream of high-energy electrons for effects on materials properties (1.4 MeV, penetration depth of electrons is about 2.5 mm). The layer of metal powder melted under the electron flow on a metal surface [13]. But in the metal all the energy of the electron flow is converted into heat. The flow of electrons is used to the direct energy transfer to target. In [14] ceramics based on magnesium fluoride doped with tungsten was first successfully synthesized using a powerful flux of electrons with energy of 1.4 MeV. The synthesis of dielectric materials in the field of a high-power flux of high-energy electrons is fundamentally different from the synthesis in thermal fields. With the used irradiation modes, which make up 20 % of the maximum possibility of the accelerator, the bulk density of the absorbed energy during the exposure to the flow for 1 s is $8 \cdot 10^{23}$ eV/cm$^3$.

During the synthesis in radiation field ~ $12 \cdot 10^{22}$ electronic excitations (ionization, electron-hole pairs) are created in 1 cm$^3$ of charge consisted of the oxide materials with an average band gap. The formation of structural phases occurs from a set of charge elements with a high degree of ionization, that is, with dangling bonds. Therefore, it is possible to form structural phases from various combinations of atoms and ions participating in the process. Since $Y_2O_3$, $Al_2O_3$, $Gd_2O_3$, $Ce_2O_3$ are dominant in the charge, the formation of the crystalline phase of YAG or YAGG with activator ions, cerium, distributed in it is most likely. During synthesis in thermal fields, only the core (atomic, ionic) subsystem is excited. The electronic subsystem remains unexcited. Unambiguous proof of the dominant role of ionization in the process of ceramics synthesis in the radiation field is the following. For the production of ceramics based on $MgF_2$, the stream with a power of 18 kW was optimal, whereas for the synthesis of YAG: Ce ceramics - 22 kW. The melting point of $MgF_2$ is 1260 K, whereas the melting points of the components for the synthesis of YAG:Ce ceramics are in range from 2060 K ($Al_2O_3$) to 2450 K ($Y_2O_3$). For the synthesis of ceramics from components, the melting points of which differ by almost 2 times, flows of similar power are needed.

Note that during the synthesis in a powerful radiation flux, a significantly higher efficiency of the mixing of charge components is achieved. This is evident from the results of the study of the spatial distribution of elements in ceramics. It should be expected that the obtained composition of ceramics will be closer to the initial one in the charge than during the synthesis by other methods.

## References

[1] Ye S, Xiao F, Pan Q, Ma Z and Zhang Q 2010 *Materials Science and Engineering reports* **71** 1–34

[2] Pan Y, Wu M and Su Q 2004 *Materials Science and Engineering B* **106** 251–56

[3] Silveira L G D., Cótica L F., Santos I A., Belançon M P, Rohling J H and Baesso M L 2012 *Materials Letters.* **89** 86–89

[4] Fadlalla M H, Tang CC, Wei S Y. and Ding X X. 2008 *J. Lumin.* **128** 1655–59

[5] Yang H.J. [et al.] 2012 *J. Am. Ceram. Soc.* **95** 49–51

[6] Liu Z, Liu S, Wang K and Luo X 2010 *Applied Optics* **49** 247–57

[7] Lisitsyn V M., Soshchin N P, Yu Yang yang, Stepanov S A, Lisitsyna L A, Tulegenova A T, and Abdullin Kh A. 2017 *Russian Physics Journal* **60** 862–65

[8] Lisitsyn V M, Abdullin Kh A, Stepanov S A, Tulegenova A T 2016 *Izv.Vuzov.Physics* **59** 164–68

[9] Lisitsyna L A, Oleshko V I., Putintseva S N. and Lisitsyn V M 2008 *Optics and Spectroscopy* **105** 531–37

[10] Lisitsyna L A, Lisitsyn V M 2013 *Physics of the Solid State* **55** 2297–2303

[11] Dolgachev G I, Kazakov E D, Kalinin Yu G, Malinin S A, Maslenikov D D and Sadovnichnii D N 2018 14th International Conference on Modification of Materials with Particle Beam and Plasma Flow *EFRE* 2018 238

[12] Dvilis E S, Burdovitsyn V A, Khasanov A O, Oks E M, Klimov A S, Zenin A A, Khasanov O L 2016 *Fundamentalnie issledovania* **10** 270–79

13] Golkovski M G, Samoylenko V V., Polakov I A, Bataev V A and Chakin I 2018 3rd International Conference on New Materials and Haigh Technologies, *EFRE* 2018 622.

[14] Lisitsyn V M, Lisitsyna L A, Dauletbekova A, Golkovskii M, Karipbayev Zh T, Musakhanov D, Akilbekov A, Zdorovets M, Kozlovskiy A and Polisadova E 2018 Nuclear Inst. and Methods in Physics Research B **435** 263–67

# The influence of light spectrum on morphogenesis of orchid germs in vitro

L V Khotskova[1], G Ya Stepanyuk[1], M S Yamburov[1], T P Astafurova[1] and S B Turanov[2]
[1]National Research Tomsk State University, Lenin Avenue 36, Tomsk, 634050, Russia
[2]National Research Tomsk Polytechnic University, Lenin Avenue 30, Tomsk, 634050, Russia

E-mail: tyrsb@yandex.ru

**Abstract.** The aim of the study is to reveal growth and developments specificities of orchid (*Cymbidium hybridum* F1) germinant *in vitro* cultured early in the ontogenesis in relation to various light spectra. Chromatic and achromatic luminous tubes (Philips, TLD 18 W) were used as lighting sources. 4 light patterns were tested: control group – colorless (white) light WL; group 1 – WL with addition of red WL+RL, group 2 – WL and blue WL+BL, group 3 – WL and green WL+GL. The authors showed that during *C. hybridum* F1cultivation in type breeding ground, morphometric growth indexes of root and leaf and wet weight of germs raised with the increase of red light part in PAR (photosynthetically active radiation) flux, thought stem length was twice bigger under the mix of colorless and green light, and also blue light addition to achromatic emission aided the reduction of germ parameters, in contrast with control group. Growth characteristics of plants, cultivated under photoculture conditions, can be adjusted by light with varied spectral content.

## 1. Introduction

In photoculture conditions growth and development of plants depends heavily on spectral and energetic characteristics of artificial light sources, which lack optimum spectral range ratio and require continual improvements [1]. Orchid family is economically valuable and leads in production sector, widely used as crop and pot culture [2-3]. However, numerous representatives of the orchid family are described as plants with slow development process; they come into reproductive period just by 7-10 years, what inhibits reproduction and cultivation with the use of traditional methods.

The goal of the research is to reveal growth and developments specificities of orchid germinant *in vitro* early in the ontogenesis in relation to various light spectra.

## 2. Equipment and Methods

The *in vitro* cultured tissue of subtropical orchid *Cymbidium* Showgirl × *Cymbidium* Lilian Stewart (*C. hybridum* F1; figure 1) was chosen as test object. Samples were provided by Siberian Botanic Garden of National Research Tomsk State University with help of reproduction [4].

IOP Conf. Series: Materials Science and Engineering **510** (2019) 012032 doi:10.1088/1757-899X/510/1/012032

**Figure 1.** Flowering *Cymbidium* Showgirl × *Cymbidium* Lilian Stewart plant in collector's fund of Siberian Botanic Garden of National Research Tomsk State University (photo by Khotskova L V 2018).

Family reproduction of orchids *in vitro* was carried out by generally recognized procedure with the use of Murashige and Skoog medium [5]. Within 15 months of cultivation *in vitro* cultured *C. hybridum* F1 germs had obtained differentiated shoots with 4-5 leaves layers and 2-3 roots, sufficient for their planting *ex vitro* in non-sterile conditions. At this growth stage of germinant morphometric parameters were measured: the length of stem, root and leaf, assimilating area on a per intact plant basis and wet germs weight. 3 containers with germs, 10 of which were then selected for analysis, were used per light settings. Statistical analysis of orchid germinant morphometric parameters was performed using MS Excel 2007 programme pack. For plants, grown under PAR (photosynthetically active radiation) light, statistical significance of differences was evaluated with use of Student's t-test ($p \leq 0.05$) and then contrasted with control group (achromatic light). Data in the pictures is arithmetical average ± average error.

As light sources for the experiments chromatic and achromatic luminescent tubes (Philips, TLD 18 W) were chosen. This lamp type is widely applied among commercial flower production and have spectrum, which covers almost whole range of PAR. Nevertheless, it is known that each PAR region (red, green or blue) separately is inapplicable for plants cultivating in protected ground conditions [1]. During the experiments 4 light patterns for seed sprouting and further cultivation were tested: control group – colorless (white) light WL; group 1 – WL with addition of red WL+RL, group 2 – WL and blue WL+BL, group 3 – WL and green WL+GL. Lighting variants were separated by lightproof screens. Light intensity of luminescent tubes was aligned to protocorm level in all experiments and remained at 3000 lux. Cultures were maintained at the temperature 23±2°C and atmosphere relative humidity 65%, 16-hour photoperiod. To evaluate and analyze irradiation facilities, PPFDs of all colour combinations at plants surface were calculated with consideration of glass container (figure 2). The measurements were conducted by applying spectrophotometer TKA-FAR (Russia, Saint-Petersburg) at the level of plants in nutriculture medium.

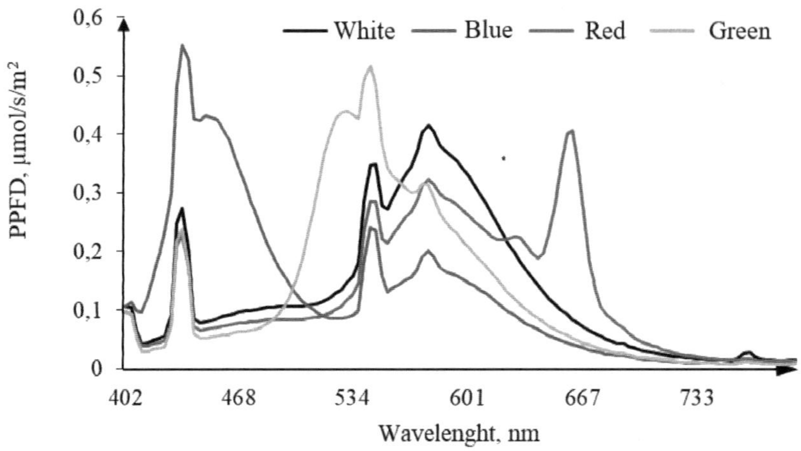


**Figure 2.** Photosynthetic photon flux density (PPFD, µmol/cm$^2$) of luminescent tubes (Philips, TLD 18 W) at the surface of *C. hybridum* plants *in vitro* cultured: control group – achromatic (white) light WL, group – WL and red light WL+RL, group 2 – WL and blue light WL+BL, group 3 – WL and green light WL+GL.

Figure 2 depicts that PAR wavelengths ratio redistributed on different sections of the plant under the flux superimposition from varied light sources. In addition, the width of peaks, peak value and spectrum ratio (BL, GL, RL) changes as well. Under mixed white and red light flux additional far-red spectrum peak appears.

## 3. Results and Discussion

It has been known, that light intensity and spectral structure affect growth and development of plants cultured *in vitro* [6-10]. In this research we detected, colorless light, combined with chromatic light, affects differently on growth and development of cultivated in typical medium orchids. Throughout continuous adaptation to varied spectral content, the combinations of white light + green or red light increased stem length of sample plants (70% and 13% higher than control group relatively), while white light + blue light lead to the degradation of stem length (figure 3). With the addition of blue light to colorless light root length of germs slightly differed from the control group, whereas in contrast the it raised by 85% and 193% over control with green and red light supplements relatively (figure 3). Total leaves area and wet weight of the germinant multiplied as well under long-wavelength spectral region (figure 4). Maximum gain of morphometric parameters occurred under WL+RL, compared with the control group.

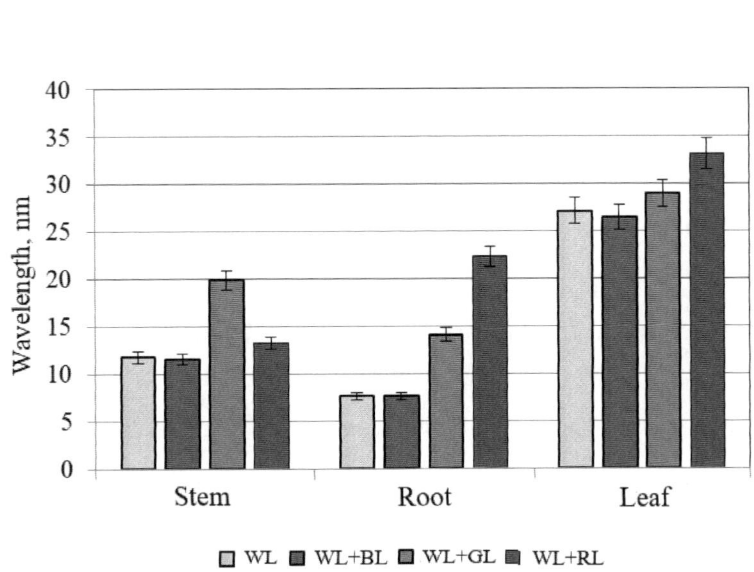


**Figure 3.** Growth parameters of 15-months *C. hybridum* F1 germs, cultivated *in vitro* in typical medium under varied light spectrum.

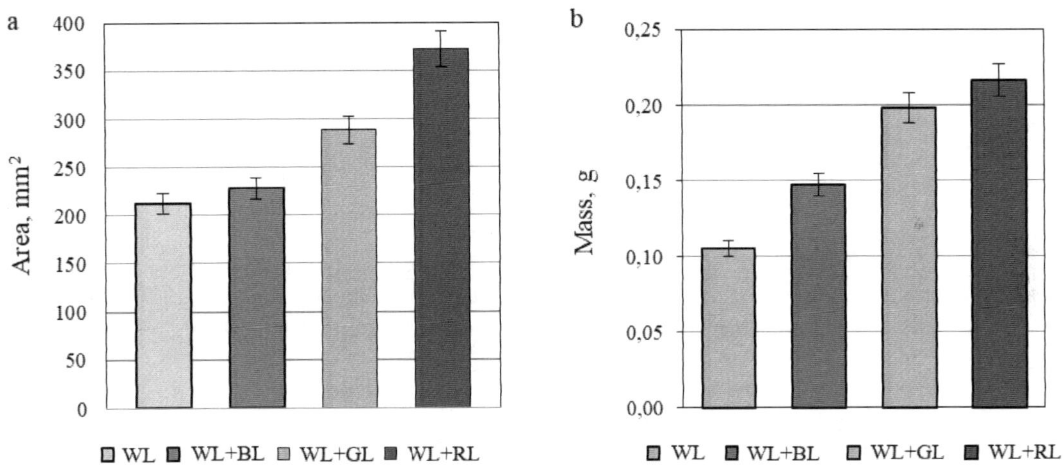


**Figure 4**. Total area (a) and wet weight (b) of the intact plant leaf (15-months *C. hybridum* F1 germinant, grown *in vitro* in typical medium under varied light spectrum).

Therefore, obtained results illustrate the relation between the development of germs *Cymbidium hybridum* F1 *in vitro* and different light spectrum early in the ontogenesis.

## 4. Conclusion

Morphometric parameters of germinant root, leaf and wet weight (grown in Murashige and Skoog medium *C. hybridum* F1 *in vitro* cultured) rose with the increase of wavelength of supplementary to achromatic light illuminating source under WL+RL and WL+GL conditions. Moreover, stem was longer under WL+GL combination. Intensified with short-wave spectral region achromatic light diminished stem length of orchid germs. Obtained results of the research ascertain, that long-wave spectral region part intensification in combines light flux enhances linear size and surface area of the plants, what was previously discovered among different species [11-14]. Under protoculture conditions growth parameters of plants can be adjusted by light emission with variously structured spectrum to modify ornamentality.

**Acknowledgement**
This research was supported by "Tomsk State University competitiveness improvement program" grant (№ 8.1.29.2018).
The work was supported by the Ministry of Education and Science of the Russian Federation under the State Program in the field of scientific investigations №13.3647.2017/PCH

**References**
[1] Bantisa F, Smirnakoub S, Ouzounisc T, Koukounarasa A, Ntagkase N and Radogloub K 2018 Current status and recent achievements in the field of horticulture with the use of light-emitting diodes (LEDs) *Scientia Horticulturae* **235** pp 437–451
[2] Kolomeitseva G L 2014 *Large-flowered orchids in the collection of Main Botanic Garden named after N V Tsitsin RAS (Cattleya, Cymbidium, Dendrobium, Paphiopedilum, Phalaenopsis)* (Moskow: GEOS) p 296
[3] De Lakshman, Pathak P, Rao A N and Rajeevan PK 2015 *Commercial orchids* ed Walter de Gruyter (India: GmbH) p 301
[4] Stepanuk G Ya and Khotskova L V 2010 The results of some tropical and subtropical orchids introduction into Siberian Botanic Garden *The first international Bekkerov readings* (Volgograd: TriAs) pp 199–201
[5] Murashige T and Skoog F 1962 A revised medium for rapid growth and bioassays with tobacco tissue culture *Physiologia Plantarum* **15** pp 473–497
[6] Kong S S, Hosakatte N M, Jeong W H, Eun J H and ,Kee Y P 2008 The effect of light quality on the growth and development of in vitro cultured Doritaenopsis plants *Acta Physiologiae Plantarum* **30** pp 339–343
[7] Rodrigues S, Joyce D, Pasqual M, Rodrigues F and de Araujo A G 2010 Etiolation and artificial light in native and hybrid orchids under in vitro cultivation *Ciencia Rural* **40** pp 1941–47
[8] Hamada K, Shimasaki K, Nishimura Y, Oyama-Egawa H and Yoshida K 2011 Effects of red, blue and yellow fluorescent films on proliferation and organogenesis in Cymbidium and Phalaenopsis PLB in vitro *Acta Horticulturae* pp 381–84
[9] Mengxi L, Zhigang X, Yang Y and Yijie F 2011 Effects of different spectral lights on Oncidium PLBs induction, proliferation, and plant regeneration *Plant Cell, Tissue and Organ Culture* **106** pp 1–10
[10] Liu H, Fu Y, Hu D, Yu J and Liu H 2018 Effect of green, yellow and purple radiation on biomass, photosynthesis, morphology and soluble sugar content of leafy lettuce via spectral wavebands "knock out" *Scientia Horticulturae* **236** pp 10–17
[11] Astafurova T, Lukash V, Goncharov A and Yurchenko V 2010 Phytotron for supplementary LED lighting of plants in greenhouses and at home *Solid-state lighting* **3** p 36–38
[12] Chung J P, Huang C Y and Dai T E 2010 Spectral effects on embryogenesis and plantlet growth of Oncidium 'Gower Ramsey' *Scientia Horticulturae* **124** pp 511–16
[13] Golovatskaya I F, Dorofeev V Yu, Medvedeva Yu V, Nikiforov P E and Karanchuk R A 2013 Optimization of illumination conditions in cultivation process of Solanum Tuberosum L. cv. Lugovskoy microcutting in vitro *Tomsk State University Journal of Biology* **24** pp 133–144
[14] Chen X L, Yang Q C, Song W P, Wang L C, Guo W Z and Xue X Z 2017 Growth and nutritional properties of lettuce affected by different alternating intervals of red and blue LED irradiation *Scientia Horticulturae* **223** pp 44–52

IOP Conf. Series: Materials Science and Engineering **510** (2019) 012033 doi:10.1088/1757-899X/510/1/012033

# Experimental results of increasing the bending strength of construction glass after microwave radiation exposure

**I V Zlobina and N V Bekrenev**
Department "Technical mechanics and machine parts", Yuri Gagarin Saratov State Technical University, 77 Politechnicheskaya street, Saratov, 410054, Russia

E-mail: irinka_7_@mail.ru

**Abstract.** The analysis of the scientific and technical literature and trends in the development of transport and energy technology showed that advanced plastic complexes and fiberglass plastics are widely used in promising aviation complexes and ground-based robotic complexes for various purposes. The promising point of application of the effects of microwave radiation (microwave electromagnetic field) on the formation of enhanced strength properties of products made of glass-reinforced plastics, especially in terms of bending stresses and interlayer shear, is shown. Experimental studies of the effect of microwave electromagnetic field with a frequency of 2450 MHz on the microstructure of hardened laminated fiberglass and their flexural strength have been performed. A 60% increase in the flexural strength of the samples was achieved after exposure to microwave radiation of an average power level for 2 minutes compared to the control ones. With shorter and longer time of processing, the increase of strength was less pronounced. The reason of hardening is an increase of the solidity of the structure by increasing the contact points of the matrix and fibers during the formation of the characteristic "star" structures. The fact of embrittlement of samples processed at maximum radiation power with an increase in exposure time was revealed.

## 1. Introduction

The development and introduction of advanced transport systems, aircrafts in particular, require the advanced creation of new high-strength and lightweight materials and the improvement of the shaping technologies for the their structural elements. Analysis of scientific and technical literature, materials of conferences and exhibitions testifies to the intensive development of the production of composite materials based on carbon-glass- and aramid fibers and their widespread use in the aviation, automotive, shipbuilding industry, rocket production and space technology in various objects. Among them are the elements of supporting structures, the plating of heavy transport vehicles, the construction of light and ultra-light unmanned and remotely piloted aircraft, including multicopters [1, 2]. The nature of the destruction of fibrous compositions under tension depends on the type and volume content of the fibers and the matrix, as well as on the ratio of their deformations to failure. The orientation of the fibers also has a decisive influence on the mechanical properties of the composition. Composites have a different mechanism for fatigue failure under cyclic loads than homogeneous materials. The complex macrostructure provides a considerably lower than that of metals, the sensitivity of composites to stress concentration and a low rate of propagation of cracks. This provides increased durability of structures made of these materials [3-7]. The use of carbon plastics is especially effective in various structural elements. According to the company, MarketsandMarkets, the global market for carbon-fiber plastics will rise to $ 35.74 billion by 2020. Carbon consumption can rise from 68 thousand tons at present to

IOP Conf. Series: Materials Science and Engineering **510** (2019) 012033 doi:10.1088/1757-899X/510/1/012033

130 thousand tons in 2020 [9]. This indicates an increased demand in the present and in the future high-tech industries for polymer composite materials.

## 2. Relevance of research

Despite the noticed increase in demand for carbon fiber, there is still a need to use composites based on fiberglass and glass fabrics. Synthetic fiberglass is widely used to create non-metallic composites - fiberglass. With a relatively low density, they have high strength, low thermal conductivity, heat, chemical and biological effects resistance. Especially important is the property of radio transparency, which determines the need to use glass-reinforced plastics in the design of radomes for radars, shells of radio receiving and radio transmitting devices of aviation robotic complexes [9, 10]. The use of fiberglass is determined here by a significantly lower tangent of dielectric loss angle compared to carbon plastics: 0.02- 0.05 and 0.17- 0.42, respectively, with a higher dielectric constant: 3.8- 8 and 3.3- 3.5. At the same time, fiberglass plastics have worse mechanical characteristics as compared with carbon plastics. In particular, the modulus of elasticity is on average 69- 80 GPa, while, as in carbon plastics, it is 120- 130 GPa with comparable tensile strength: 1920 and 1800 MPa, respectively [11- 13]. The requirement of uniform strength of the structure, for example, an airframe of an aircraft, necessitates the reinforcement of fiberglass elements, which is carried out, as a rule, by constructive methods by increasing the layers and changing the reinforcement scheme. However, this is accompanied by an increase in the weight of the structure, which is an undesirable factor for aircraft technology. Weight growth is also associated with a large proportion of fiberglass as compared with carbon plastic [11].

Consequently, one of the main tasks of creating promising technology and, mainly, aviation, is to develop design and technological methods to improve the strength characteristics of parts made of structural glass-reinforced plastics without significantly increasing the weight of the structure.

## 3. Problem statement

Recently, alternative methods have been developed to increase the strength properties of polymer composite materials and, in particular, glass-reinforced plastics. A method of creating so-called self-reinforced plastics is being developed, according to which the product is pressed from glass fibers without the use of a binder and heated to temperatures that cause activation of the micro melting of the surface of the fibers. In this case, their connection is carried out by bonding the fibers through this molten layer, which replaces the matrix of ordinary PCMs. This reduces the weight of the composition, since there is no additional matrix material, and the strength even rises a little [14]. However, this method is difficult to implement due to the complex controllability and uniformity of heating of the peripheral zone of the fibers. The change in existing technology and equipment is required. The "technological heredity" of subsequent operations and transitions influences the final set of properties. Obviously, it is more expedient to impact the final product made of hardened material.

For local control of the structure and strength properties of a three-dimensional or two-dimensional object of non-metallic materials one of the most effective methods is the use of microwave radiation (microwave electromagnetic) field. However, the majority of studies of these electrotechnological processes concerned the effects on the process of synthesizing components or curing the composition [15- 17]. Practically there is no information about the mechanism of the effect of microwave radiation on the cured structure of the finally formed product. At the same time, the use of this impact on composite materials in the process of their formation is not always advisable because of the need to make changes in the streamlined and complex technological process of component synthesis, laying out the composite layers, hardening and finishing. Previously we have obtained positive results on microwave hardening of pultruded carbon and layered carbon plastics [18, 19]. However, the influence of microwave radiation on fiberglass requires additional study due to the substantial dependence of the process of interaction of microwave radiation with PCM on their composition, including the type of fibers and the matrix material.

IOP Conf. Series: Materials Science and Engineering **510** (2019) 012033 doi:10.1088/1757-899X/510/1/012033

The aim of our research is to determine experimentally the possibility of strengthening microwave modification of the finally formed and hardened structure of structural composite fiberglass in modes that do not lead to material heating above 35- 40 °C.

## 4. Research methods and equipment

We carried out studies of the effect of microwave processing on the physicomechanical and protective properties of samples formed from 24 layers of aramid fiber TSVM-DJ article 56319A with a binder- Glue 88. While preparing samples, each layer was impregnated with a layer of glue, placed under a press and held for 24 hours after the composition was cured, one part of the samples was left as a control, and another part was placed in a microwave electromagnetic field of average power density for 3 minutes. The treatment was carried out using the installation "Zhuk-2-02" with a horn radiating antenna manufactured by LLC AgroEkoTech (Obninsk, Kaluga region). The power of the installation magnetron is 1200 W, the frequency is 2450 MHz. The distance to the plane of the aperture of the horn was 150 mm, which corresponds to the average power level. According to our earlier studies [9], it is the average microwave power that provides the greatest efficiency in increasing the bending stresses of aramid filaments. Three specimens were made with dimensions of 60 x 60 mm (Figure 1).

## 5. Theoretical part

At present, the mechanism of microwave radiation on polymeric and composite materials is described as follows [15, 17, 20, 21]. The standard frequency for microwave ovens and industrial microwave installations is 2450 MHz. This is the frequency of resonance absorption for water molecules. Under the influence of physical vibrational effects, as with heating, the mobility of the structural elements of the polymer increases. It means that such effects are equivalent to an increase in temperature. The greatest effect from the use of vibrational effects should be expected when the frequency of external exposure coincides with the natural frequency of the structural elements of the macrochains [20]. When using microwave radiation at the stage of hardening, there is an intensification of diffusion processes that promote the formation of bonds in a shorter period of time and provide greater strength of interaction between the components.

On the other hand, the performed analysis suggests that oscillatory processes in the microstructure of solidified PCMs can lead to its restructuring, changes in the number and nature of mechanical and intermolecular bonds in the matrix-fiber system and, as a result, lead to an increase in local and general strength properties subject to the avoidance of thermal destruction.

## 6. Results and discussion

We have investigated samples of laminated glass-reinforced plastics in the form of plates with dimensions of 80.0 x 15.0 x 2.0 mm. In the experiments, the microwave setting "Zhuk-2-02" produced by "AgroEkoTeh ', Obninsk, Kaluga Region, with radiating antenna horn type was used. The device generates an electromagnetic field of 2450 MHz with an output power of 1200 W. Three modes of microwave power were used: low $P_I$, medium $P_{II}$, and high $P_{III}$. Because of the patenting of the method of modifying at the present time, specific technological regimes are not indicated. The processing time was set equal to 1, 2, 3 minutes at different power levels. Tests of samples before and after processing were carried out on an installation equipped with strain gauge force sensors and a worm loading mechanism. Processing the results of the measurement of the load applied to the sample was carried out using a special program installed in the program (LabVIEW, Orel) and allowed to obtain load graphs (torque on the drive) in the dynamics from the moment of application to failure. According to known dependencies, the mechanics of materials calculated the bending stresses. The surface of the deformation zone was studied using a Digital Microscope 2.0 MP 1000X digital microscope (GAOSUO, China) at x500 magnification with display on a laptop screen.

According to the results of processing the experimental setup, graphical dependencies were obtained (Figure 1).

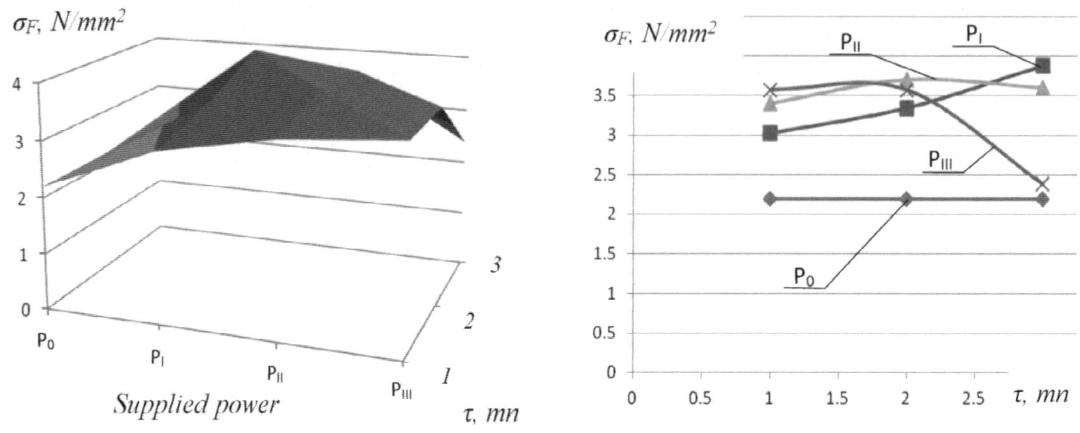


a b

**Figure 1.** The effect of microwave processing time on fiberglass on bending stresses σF at different supplied power

With a short exposure time, a steady increase in the specimen strength is observed with an increase in the input radiation power, which is determined by the limiting bending stresses; with an increase in time, the specimen strength increases, but tends to decrease with increasing power. Finally, at the maximum of the investigated range of exposure time and low power, the maximum hardening effect is achieved. The effect of hardening on bending stresses in this case is more than 60 % and exceeds that previously achieved for CFRP. The hardening effect at medium power decreases when the exposure time exceeds the level of 2 minutes. At the maximum of the investigated power range, the hardening effect sharply decreases with increasing time. The strength of the sample remains almost comparable with the control. The study of the microstructure of the sample deformation zone (Figure 2) showed the following.

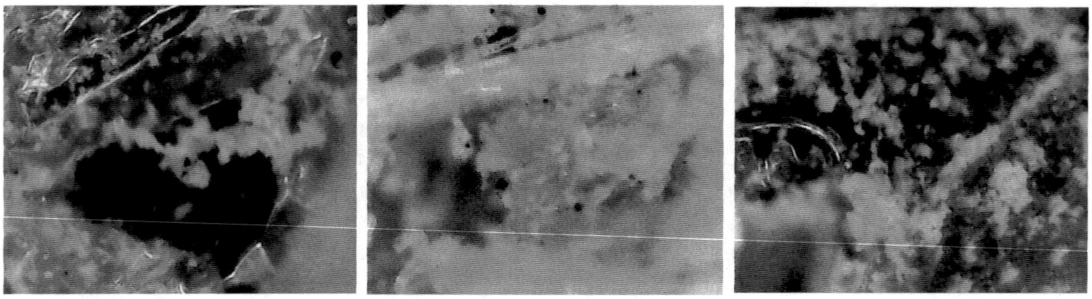

a b c

**Figure 2.** The microstructure of the control sample of fiberglass (a) and processed with medium (b) and high (c) microwave power x500

In the fracture zone of the control sample (Figure 2a), fiber breaks and particles of the crumbling matrix are visible. The structure of the sample processed at low power is characterized by solidity; the fibers are interconnected by "star-shaped" matrix structures having a developed contact surface. When processed at medium power for 2 minutes (Figure 2b), individual fibers are visible, separated from each other, although with preserved "star-like" formations on the surface. After exposure for 3 minutes at high power, the sample looks completely destructed (Figure 2c). The fibers are separated; the particles of the matrix are split and are not connected to each other. In general, this structure resembles the structure of control samples.

The results obtained can be explained as follows. Over time, at a low power level, the processes of formation of additional cross-links gradually develop, both in the matrix and between the matrix and the reinforcing fibers, as well as between the fibers themselves, which form conglomerates (welded strands). Due to the formation of such a maximally connected structure, the material acquires maximum strength. With increasing radiation power supplied to fiberglass, the heat generation increases, which on the one hand accelerates the hardening process (maximum occurs after 2, not 3 minutes), on the other hand, as the exposure time increases, it leads to dehydration and embrittlement of both the matrix and the reinforcing fibers, which manifests itself in reducing strength. At maximum power, this mechanism is manifested to a greater degree; therefore, an increase in the exposure time causes heating of the material, which adversely affects the adhesion of the components and also the cohesive interaction of the structural formations of the matrix. Accordingly, the resistance to the action of loads decreases and reaches a minimum during long-term processing at the maximum of the investigated power range.

## 7. Conclusion

It was experimentally shown that the effect of microwave radiation on structural elements of hardened fiberglass for 1-3 minutes allows to increase the strength of bending stresses by 60 %. At the same time, the greatest effect is achieved when processing with a minimum power level for 3 minutes or with an average level for 2 minutes.

The reason for the increase in strength may be an increase in the points of interaction of the fibers with the matrix and the fibers between themselves (with the formation of compacted bundles and "star" structures) due to local thermal effects in the interfacial zones.

**Acknowledgments**
The studies were carried out with the support of the RFBR grant No. 17-03-00720 “Methodology of optimization micro-design of composite materials for objects of complex shape with increased dynamic strength, which are layer-by-layer formed by electrotechnological methods”.

## References

[1] Koshkin R 2015 *Scientific Bulletin of the State Civil Aviation Institute of Civil Aviation* vol. 10 pp 16–25
[2] Kiva D 2014 *Aerospace Engineering and Technology* **6** 5
[3] Kablov E 2012 *Bulletin of the Russian Academy of Sciences* **82** 520
[4] Ahmad A, Abdalla M and Gurdal Z 2010 *Journal of Aircraft* **47** 775
[5] Brinkmann S at al. 2006 International Plastics Handbook the Resource for Plastics Engineers *Ed. Hanser* p 920
[6] Bunakov V, Vasiliev V and Gurdal Z 1999 *Optimal Structural Design USA : Technomic Publishing* pp 207–246
[7] Karbhari V et al. 1998 *ECCM-8 European conference on composite materials «Science, technologies and applications»* **2** 35
[8] Avrasin A, Borodin M and Kiselev B 1982 *Aviation industry* **8** 80
[9] Preobrazhensky A 2010 *Chief Mechanic* **5** 27
[10] Vasilyev V, Protasov V, Bolotin V and others 1990 Composite materials: Reference book *Moscow Mashinostroenie* p 512
[11] Kurnosov A, Melnikov D and Sokolov I 2015 *Proceedings of VIAM* **8** 08
[12] Davydova I and Kavun N 2012 *Glass and Ceramics* **4** 1–7
[13] Sevostyanov D, Doriomedov M, Daskovsky M et al. 2017 *Proceedings of VIAM* **4** 104

[14] Arkhangelskiy Y 2011 The reference book on the microwave electrothermies: a reference book *Saratov: A scientific book* p 560
[15] Kolomeitsev V, Kuzmin Y, Nikuiko D and Semenov A 2013 *Electromagnetic Waves and Electronic Systems* **18** **12** 25
[16] Estel L, Lebaudy Ph, Ledoux A, Bonnet C and Delmotte M 2004 *Proceedings of the Fourth World Congress on Microwave and Radio Frequency Applications* **11** 33
[17] Zlobina I, Bekrenev N and Muldasheva G 2016 *International Conference on Advanced Materials with Hierarchical Structure for New Technologies and Reliable Structures* **1783** 020236-1
[18] Zlobina I, Bekrenev N and Pavlov S 2017 *Bulletin of the Chuvash State Pedagogical University. AND I. Yakovlev. - Ser .: Mechanics of ultimate state* **3** 42
[19] Komarov V 2010 *Physics of wave processes and radio engineering systems* **13** **4** 57
[20] Studentsov V and Pyataev I 2014 Effect *Russian Journal of Applied Chemistry* **87** **3** 352